Berend Eling, Wolfgang Friederichs
Polyurethanes

Also of interest

Polyurethane Foams.
Coconut Oil-Based Polyols
Arnold A. Lubguban, Gerard G. Dumancas, Roberto M. Malaluan and Arnold C. Alguno, 2025
ISBN 978-3-11-163264-3; e-ISBN (PDF) 978-3-11-163266-7

Two-Component Polyurethane Systems.
Innovative Processing Methods
Chris Defonseka, 2019
ISBN 978-3-11-063957-5; e-ISBN 978-3-11-064316-9

Polyurethanes.
Coatings, Adhesives and Sealants
Hans-Ulrich Meier-Westhues, Karsten Danielmeier, Peter Kruppa and Edward Squiller, 2019
e-ISBN (PDF) 978-3-7486-0047-3

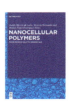
Nanocellular Polymers.
From Microscale to Nanoscale
Miguel Angel Rodríguez Pérez, Judith Martín de León and Victoria Bernardo García, 2024
ISBN: 978-3-11-075611-1; e-ISBN 978-3-11-075613-5

Polymer Matrix Composite Materials.
Structural and Functional Applications
Debdatta Ratna and Bikash Chandra Chakraborty, 2023
ISBN 978-3-11-078148-9, e-ISBN (PDF) 978-3-11-078157-1

Berend Eling, Wolfgang Friederichs

Polyurethanes

Polyols, Isocyanates, Rigid Foams, Flexible Foams, Elastomers

DE GRUYTER

Authors
Prof. Dr. Berend Eling
University of Hamburg
Institute of Technical and Macromolecular Chemistry
Bundesstrasse 45
20146 Hamburg
berend.eling@uni-hamburg.de

Dr. Wolfgang Friederichs
Am Ackerrain 16
50933 Köln
Germany
wf@purecat.eu

ISBN 978-3-11-074456-9
e-ISBN (PDF) 978-3-11-074458-3
e-ISBN (EPUB) 978-3-11-074462-0

Library of Congress Control Number: 2025937528

Bibliographic information published by the Deutsche Nationalbibliothek
The Deutsche Nationalbibliothek lists this publication in the Deutsche Nationalbibliografie;
detailed bibliographic data are available on the Internet at http://dnb.dnb.de.

© 2025 Walter de Gruyter GmbH, Berlin/Boston, Genthiner Straße 13, 10785 Berlin
Cover image: BASF Polyurethanes GmbH
Typesetting: Integra Software Services Pvt. Ltd.

www.degruyter.com
Questions about General Product Safety Regulation:
productsafety@degruyterbrill.com

Preface

Polyurethanes are well-established materials. With an annual production volume of 25 million metric tons, they belong to the commercially most important specialty polymers. Polyurethanes are widely used as flexible foam for furniture, rigid foam for insulation, and various elastomer applications such as shoe soles and steering wheels. The manufacturing of polyurethanes from liquid reactive components enables the production of low-density cellular materials, and the wide variety of starting components allows for a broad spectrum of properties, ranging from rigid and glassy to soft and elastomeric. The vast spectrum of polyurethanes and the large variety of available starting materials introduce a high level of complexity, requiring a basic fundamental understanding of their structure-property relationships.

A master course at the University of Hamburg, Institute of Technical and Macromolecular Chemistry, Germany, 2010, aimed to teach the basics of polyurethanes. The present textbook is an extended version of this course. It is especially suited for master courses at universities and polytechnics, as well as for newcomers working in research and development within the polyurethanes industry.

This textbook approaches the subject from an application viewpoint. It provides a fundamental understanding of the required basic organic, physical, and polymer chemistry and links the starting materials to the polymer morphology and mechanical properties of polyurethanes. Keeping the book down to a manageable size, we concentrated on the three main applications: low-density rigid and flexible foams and elastomers.

Our special thanks go to Dr. Günter Scholz for bringing us together and establishing contact with the publisher. We thank Prof. Almut Stribeck and Dr. Mengyu Zhang for critically reading parts of the manuscript. We are very much indebted to BASF Polyurethanes and Covestro for allowing us to present some data on physical properties. Sincere thanks are given to AutoRIM Ltd, BASF-Polyurethanes, Bucher Hydraulics GmbH, Cannon S.p.A., Covestro AG, DESMA Schuhmaschinen GmbH, Graco Inc., Hennecke GmbH, KraussMaffei Technologies GmbH, Ottobock SE & Co. KGaA, Prüfinstitut Hoch, puren GmbH, SPFA (Spray Polyurethane Foam Alliance), Soudal N.V., and SATRA Technology for providing pictures that illustrate the processing, properties, and polyurethane applications.

Spring 2025

Berend Eling
Wolfgang Friederichs

Contents

Preface —— V

1 Introduction —— 1
1.1   Historical background —— **3**
1.2   Polyurethane market —— **5**
       References —— **8**

2 Starting components —— 9
2.1   Polyisocyanates —— **9**
2.2   Polyether polyols —— **23**
2.3   Polyester polyols —— **32**
2.4   Bio-based polyols —— **35**
2.5   Hydroxyl value and equivalent mass —— **37**
2.6   Diamines and polyamines —— **38**
2.7   Additives —— **40**
       References —— **46**

3 Polyurethane chemistry —— 47
3.1   Reactivity of the isocyanate group —— **47**
3.2   Isocyanate reactions with active hydrogen compounds —— **49**
3.3   Isocyanate-isocyanate reactions —— **58**
3.4   PU system technology —— **63**
3.5   Chain topology and polymer morphology —— **66**
3.6   Rheology and cure —— **71**
3.7   Structure development and reaction rate —— **73**
       References —— **79**

4 Physical properties and flammability —— 81
4.1   General thermal behavior of polymers —— **81**
4.2   Viscoelasticity —— **84**
4.3   Dynamic mechanical analysis —— **87**
4.4   Melting of the hard domains —— **89**
4.5   Burning behavior and flame protection —— **93**
       References —— **97**

5 Processing —— 98
5.1   Prepolymer process and one-shot method —— **98**
5.2   Discontinuous and continuous processing —— **98**
5.3   Low-pressure processing —— **100**
5.4   High-pressure processing —— **101**

VIII —— Contents

5.5	RIM technology —— **102**	
5.6	Equipment —— **104**	
	References —— **110**	

6 Foam formation —— 111

6.1	Simultaneous formation of polymer and foam —— **112**
6.2	Aeration and nucleation —— **113**
6.3	Bubble growth —— **115**
6.4	Fine-cell rigid foams —— **122**
6.5	Foam properties —— **123**
	References —— **124**

7 Rigid foams —— 125

7.1	Rigid foam formulations —— **125**
7.2	Glass transition temperature —— **129**
7.3	Foam formation —— **131**
7.4	Properties —— **134**
7.5	Applications and processing —— **147**
	References —— **157**

8 Flexible foams —— 159

8.1	Foam properties —— **161**
8.2	Flexible foam formulations —— **164**
8.3	Polymer topology and morphology —— **166**
8.4	Manufacturing of open-cell foam —— **169**
8.5	Morphology and polymer hardness —— **174**
8.6	Compression hardness —— **176**
8.7	Ball rebound resilience, hysteresis, and loss factor —— **179**
8.8	MDI versus TDI technology —— **179**
8.9	Processing —— **181**
8.10	Technical foams —— **185**
	References —— **188**

9 Elastomers —— 190

9.1	Elastomer starting materials and formulations —— **191**
9.2	Chain topology —— **194**
9.3	Polymer morphology —— **196**
9.4	Effect of processing on morphology —— **208**
9.5	Performance-related tests —— **212**
9.6	Mechanical properties —— **214**
9.7	Applications —— **218**
	References —— **222**

Contents —— **IX**

10 Sustainability and outlook —— 224
10.1   Improved carbon footprint of PU starting materials —— **224**
10.2   Emission and odor —— **225**
10.3   Insulation —— **226**
10.4   Recycling of PU —— **226**
10.5   Epilogue —— **227**
        References —— **228**

11 Symbols and abbreviations —— 229

12 Unit conversion tables —— 233

13 Calculations —— 236
13.1   Basic equations —— **236**
13.2   Worked examples —— **237**
13.3   Average functionality of a polyol mixture —— **238**
13.4   Hard block content of an elastomer —— **238**

Index —— 241

1 Introduction

Polyurethanes are found in a wide range of forms and applications. They are used as flexible foam for cushioning applications, as rigid thermal insulation materials, or as elastomers with a soft touch, such as in shoe soles and steering wheels. Polyurethane, abbreviated as PU, is one of the most commercially important specialty polymers, with an annual global consumption of approximately 25 million metric tons.

The term polyurethane is used to designate polymers produced by the polyaddition of polyfunctional isocyanates with molecules that bear at least two hydroxyl groups [Fig. 1.1]. The name polyurethane is derived from the urethane group formed in this reaction. More generally, one can say that isocyanate reacts with groups that contain "active" hydrogen atoms, such as OH and NH, like alcohols, carboxylic acids, amines, and water. PU materials may, therefore, also contain other linkages, such as urea and amide.

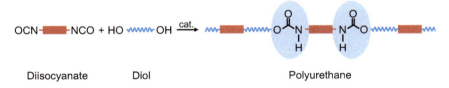

Fig. 1.1: Polyaddition reaction of a diisocyanate and a diol to form a polyurethane.

Diphenylmethane diisocyanate (MDI) and toluene diisocyanate (TDI) are the most commonly used isocyanates. The isocyanate groups in both isocyanates are linked to an aromatic ring and concomitantly activated by resonance stabilization. The intrinsic activation of the aromatic isocyanate ensures that the polymerization reaction takes place under mild conditions. The most prominent reaction partners of isocyanates are polyether- and polyester-based polyols. The polyols are telechelic, meaning the hydroxyl groups are positioned at the chain ends. The second most important reaction is that of water and isocyanate, yielding a urea bond for building the chain and carbon dioxide to expand the foam. Water is, therefore, also referred to as a chemical blowing agent. Catalysts are required to accelerate the reaction and to control the selectivity of the different isocyanate reactions during polymerization. Physical blowing agents, such as pentane, can also be used to expand the foam. Surfactants stabilize foam growth, and flame retardants can be added to reduce flammability.

The components used for the production of polyurethanes are, in general, low-viscosity liquids. Polymerization requires intensive mixing of the polyol and isocyanate components using dedicated PU processing machines. The number of isocyanate-building blocks on the industrial megaton scale is limited, but the choice of polyols and additives is abundant. The wide range of components enables the production of tailor-made polymers for various applications according to customer specifications.

The polyurethane reaction between polyol and polyisocyanate is a step-growth polymerization: all hydroxyl and isocyanate groups in the system can react, and the chains grow until the reaction is complete. A key feature of the step-growth reaction is that the viscosity gradually increases. This gives polyurethanes two significant benefits that other polymers do not possess: they are strong adhesives and can easily be expanded, providing low-density foams.

The carbon dioxide generated in the urea-forming reaction expands the foam. The liquid resin viscosity builds up slowly, allowing the foam to expand at low pressures. The generation of low pressures during foam expansion is fundamental to achieving low foam densities and excellent foam flow. Low-density foams can be produced, and complex mold geometries can be filled with PU foam. The density of polyurethanes can be varied over two orders of magnitude, ranging from 1,200 to 10 kg/m^3.

During the reaction, polyurethanes go through a tacky phase. When the tacky polymer comes into contact with a substrate, it provides excellent wetting, and after curing, it forms a strong adhesive bond. This enables the facile production of composite materials, such as aluminum-foam-aluminum panels for building applications or plastic-foam-aluminum materials for appliances like freezers and refrigerators. Instrument panels for the automotive industry can be produced in one production step. The polyurethane foam offers excellent flow, enabling the complete filling of the narrow cavity between the insert and decorative foil. Upon full cure, it provides excellent adhesion to both substrates, resulting in a strong, load-bearing automotive part.

The choice of polyol largely determines the mechanical properties of PU polymers. The polyols are characterized by their molar mass and the number of hydroxyl groups per molecule, known as the hydroxyl functionality. The hydroxyl equivalent mass is calculated by dividing the molar mass by the hydroxyl functionality. Polyols with a hydroxyl functionality ranging from 3 to 8 and an equivalent mass of 100 – 200 g/mol produce rigid, glassy materials. In contrast, polyols with a hydroxyl functionality of 2 – 3 and an equivalent mass between 500 and 2,000 g/mol yield soft elastomers. By varying the polyol, the polymer modulus of polyurethanes can be altered over three orders of magnitude, ranging from approximately 1 MPa to 1 GPa. PU polymers are typically produced from branched reactants, resulting in chemically crosslinked polymer networks. Thermosets are produced in their final form and shape, such as a molded seat for automotive applications.

Depending on the foam recipe, open-cell and closed-cell foams can be produced. Flexible foams have an open-cell structure. During the last stages of the foam expansion, the cell membranes rupture, and the polymer material retracts to form a three-dimensional array of elastic struts. This open elastic structure provides a soft touch and excellent cushioning properties. Rigid foams are closed-cell materials with excellent thermal insulation properties because the enclosed gases in the cells exhibit lower conductivities than air.

Polyurethanes can be produced in a wide range of grades, addressing applications in automotive, furniture and bedding, construction, thermal insulation, and footwear. All PU applications can be mapped in a hardness-density plot, as shown in Fig. 1.2. Three foam types are significant in production volumes: low-density flexible foam, low-density rigid foam, and high-density elastomers.

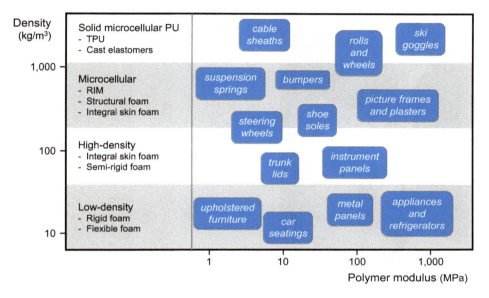

Fig. 1.2: Hardness-density matrix of polyurethanes showcasing the main applications.

1.1 Historical background

A new man-made fiber, manufactured and patented by DuPont, was introduced to the market in the mid-1930s. The fibers were spun from Nylon 66, the polycondensation product of adipic acid and 1,6-hexanediamine. The German company IG Farbenindustrie produced a semi-synthetic fiber, based on cellulose, known as rayon. They sought to overcome the threat posed by novel nylon fibers through the work of Otto Bayer and his team, who developed the polycondensation process of diisocyanates with diols and diamines. A patent was filed in 1937 (DE 728981) claiming a "process for the production of polyurethanes and polyureas, characterized by reacting organic diisocyanates with compounds bearing at least two hydroxyl or amine groups with exchangeable hydrogen atoms". The polyurethane business of IG Farbenindustrie in Leverkusen was transferred to Bayer after the Second World War, and in 2015, it became the core business of Covestro.

Spinning fibers from the newly invented polyurethanes, however, proved to be problematic. The fibers fractured when gas bubbles appeared in the spinning nozzle.

Otto Bayer discovered that these gas bubbles were caused by carbon dioxide generated in the reaction of the isocyanate with traces of water in the polyol. He took advantage of this problem and added more water to the reaction system – and PU foams were invented [Fig. 1.3].

Fig. 1.3: Prof. Otto Bayer demonstrates the preparation of polyurethane foam (with the kind permission of Covestro AG).

After the war, the military government of the Allied Forces ordered the dismantling of several IG Farbenindustrie production plants. The production of TDI was permitted and restarted again in 1948. The first industrial application of PU foam was a twin-walled beer keg insulated with rigid foam based on TDI and polyester polyol, with an annual production volume of 18 tons.

Hennecke invented the high-pressure mixing process of the reactants. The mixing process was described as "conveying isocyanate and polyol by piston pumps and introducing them via injection nozzles into a mixing chamber". The high-pressure processing mixing head, with an annual output of approximately 360 tons, was introduced in 1953.

One year later, in 1954, the polyurethane market was expanded to include the USA. Mobay, a joint venture between Bayer and Monsanto, began producing polyols and isocyanates for the American PU market. Dow, possessing propylene oxide technology, started producing polyether polyols for use in polyurethane applications. Their polyether polyol plant in Freeport came on stream in 1959. The Asian market was entered with local production of PU sponges and mattresses. In the 1960s, the global PU market volume surpassed the 100 tons per year mark.

The high versatility of PU allowed the development of several high-volume applications. The flexible foam market for bedding and furniture was established in the 1960s, with applications in thermal insulation, automotive, construction, and footwear following shortly thereafter. The global PU market had grown to approximately 2.7 Mio. mt/y (million metric tons per year) in 1980, demonstrating the enormous potential of PU.

1.2 Polyurethane market

From 1980 to the present, the market has increased by a factor of ten [Fig. 1.4], corresponding to an average annual growth rate of 5.6% over more than 40 years.

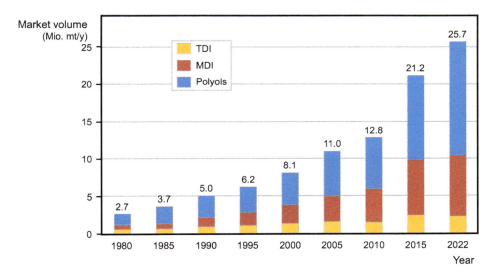

Fig. 1.4: Global polyurethane market in Mio. mt/y, subdivided into TDI, MDI, and polyol [1].

The global PU market volume in 2022 was 25.7 Mio. mt/y. This may sound impressive, but PU is considered a niche polymer, with a market share of approximately 5.5% [Fig. 1.5]. The largest volume polymers are the low-cost thermoplastics polyethylene (PE), polypropylene (PP), and polyvinyl chloride (PVC), with market shares of 26.9, 19.3, and 12.9%, respectively. PU ranks fifth in volume and is the second most popular among specialty polymers, after polyethylene terephthalate (PET).

Asia Pacific dominates today's global PU foam market [Fig. 1.6], primarily driven by China, which accounts for a 39% market share by value. Asia Pacific is expected to maintain its leading market position in the future, as steadily increasing wealth in the region is anticipated to drive the demand for PU. The markets of North America and Europe account for approximately half of the global PU business, with individual

6 — 1 Introduction

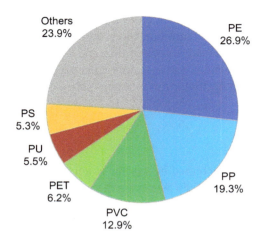

Fig. 1.5: Global plastics market in percent [1].

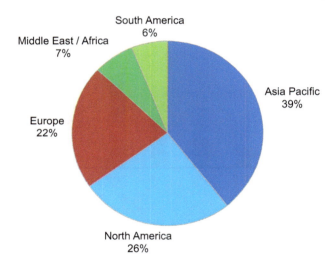

Fig. 1.6: Global polyurethane foam market 2022 by region and value [1].

contributions of 26 and 22%, respectively. The Middle East, Africa, and South America collectively have a market share of approximately 13%.

When considering the PU market by end-use application [Fig. 1.7], the primary application areas are construction (predominantly rigid foam) and furniture (flexible foam), with volume shares of 26 and 22%, respectively. These two sectors are followed by automotive (17%) and electronics and appliances (14%). Appliances use rigid foam, whereas the automotive and electronics sectors (17%) show a broad range of PU products, including various grades of flexible foams and elastomers. The footwear sector

(5%) and packaging sector (4%) include elastomers and flexible foam, respectively. These six main application areas account for nearly 87% of the polyurethane market. The remaining 13% comprises a variety of smaller-volume products, including adhesives, coatings, sealants, and binders.

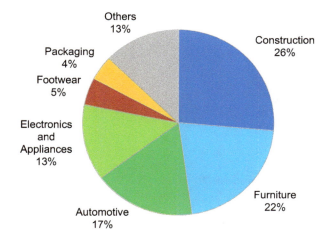

Fig. 1.7: Global polyurethane market 2022 by applications [1].

The PU industry began with TDI and polyester polyols; however, today's market is dominated by MDI and polyether polyols [Fig. 1.8]. The production volumes and growth rates of TDI and MDI were comparable until approximately 1994; thereafter, the growth rate of the MDI market accelerated, and it soon became the dominant isocyanate. The strong growth of MDI was due to its traditional markets, notably rigid foam, experiencing rapid growth, as well as the discovery of new markets for MDI, such as its use as a wood binder, replacing phenol-formaldehyde resins. Moreover, MDI started substituting TDI, for instance, in molded flexible foam for automotive seats and office furniture. In 2020, the production volume of MDI (8.0 Mio. mt/y) was approximately three times that of TDI (2.5 Mio. mt/y). Aliphatic isocyanates play only a minor role, with an estimated market volume of approximately 4% (0.5 Mio. mt/y) of the total isocyanate production.

The first PU applications were polyester-based, but this changed in the late 1950s, primarily due to the flexible foam industry's shift to polyether polyols. In 2020, the production volume of polyether polyols (9.6 Mio. mt/y) was about 78% of the total polyol produced, and that of polyester polyols was about 22% (2.6 Mio. mt/y).

1 Introduction

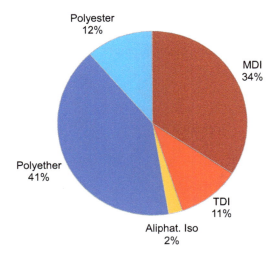

Fig. 1.8: Global polyurethane market 2020 by raw materials [2].

References

[1] The presented market data are averaged data taken from "Polyurethane Foam Market by Type, End-use Industry, and Region – Global Forecast to 2028" (https://www.marketsandmarkets.com/Market-Reports/polyurethane-foams-market-1251.html) "Global Polyurethane (PUR) Foam Market to Reach a CAGR of 6.0% from 2020 to 2028"; Press release from Quince Market Insights (https://www.openpr.com/news/2297703/global-polyurethane-pu-foam-market-to-reach-a-cagr-of-6-0-from); "Polyurethane Foam Market Share, Size, Trends, By Regions; 2021–2028 (https://www.polarismarketresearch.com/industry-analysis/polyurethane-foam-market); "Industrial Foam Market by foam type, by end-use industry, by resin type – Global Industry Analysis; Regional outlook and forecast 2023–2032 (https://www.precedenceresearch.com/industrial-foam-market); "Polyurethane Market Size, Share & Trend Analysis 2023–2030" (https://www.grandviewresearch.com/industry-analysis/polyurethane-pu-market); "Polyurethane Market by Product and Application; Global Industry Analysis 2023–2032" (https://www.precedenceresearch.com/polyurethane-market); "Polyurethane Market – Global Industry Analysis and Forecast (2023–2029)" (https://www.maximizemarketresearch.com/market-report/polyurethane-market/70607/).

[2] A. Austin, D.A. Hicks, A review of the global PU industry of 2015 and outlook for 2016, PU Magazine, Vol. 13, No.1, 22–34.

2 Starting components

Polyisocyanates and polyols are the primary starting materials for producing polyurethanes. The isocyanates are subdivided into aromatic and aliphatic isocyanates, and the polyols into polyether and polyester polyols. The most important isocyanates are MDI and TDI. The structure of polyols can be easily varied, and numerous polyols are available for various PU applications. Additives create specific effects; for instance, blowing agents and surfactants are required to expand the foam, and fire retardants are used to improve fire resistance.

2.1 Polyisocyanates

C.A. Wurtz was the first to describe an isocyanate formed by the reaction of dimethyl sulfate with potassium cyanate in 1848.

Other routes to isocyanates are nitrene rearrangements, known as Hofmann, Curtius, and Lossen rearrangements. These synthesis methods have been of academic interest only so far. For the industrial manufacturing of isocyanates, the phosgenation of amines has become the state-of-the-art process.

Polyurethane technology uses both aromatic and aliphatic polyisocyanates. Aromatic polyisocyanates exhibit higher reactivity toward polyols, and their rigid structure typically yields polyurethanes with superior mechanical properties. Furthermore, their manufacturing is more cost-efficient.

Methylene diphenyl diisocyanate (MDI) and toluene diisocyanate (TDI) are by far the most important isocyanates for producing polyurethanes, accounting for more than 95% of the worldwide isocyanate production. Other aromatic isocyanates, mainly used in high-performance elastomer applications, are 1,5-naphthalene diisocyanate (NDI), p-phenylene diisocyanate (PPDI), and 3,3'-dimethyl-4,4'-biphenylene diisocyanate ("o-tolidine diisocyanate"; TODI).

Aliphatic isocyanates only play a subordinate role in the PU industry. They are primarily used in the coatings industry because of their non-yellowing properties. The major contenders of the aliphatic isocyanates are hexamethylene diisocyanate (HDI) and isophorone diisocyanate (IPDI). Furthermore, hydrogenated MDI (H_{12}MDI) and tetramethylxylylendiisocyanat (TMXDI) are used. Bio-based pentamethylene diisocyanate (PDI) became commercially available in the last decade.

https://doi.org/10.1515/9783110744583-002

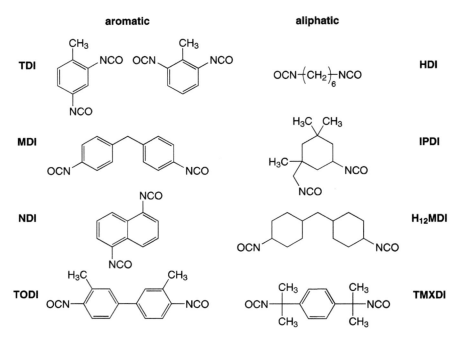

The synthesis of isocyanates typically begins with hydrocarbon precursors, such as benzene or toluene. The precursor is first nitrated and subsequently hydrated to the corresponding amine. The amine, in turn, is phosgenated, yielding the crude isocyanate. The crude isocyanate is then purified and optionally fractionated into differentiated product streams. The precursor structure differences may affect the introduction of the amine groups; the amine phosgenation reaction for all isocyanates is essentially the same.

The phosgenation of amine to isocyanate is a two-step reaction in which carbamoyl chloride is formed first. Hentschel discovered this synthesis route to isocyanate in 1884.

Phosgene, formally the acyl chloride of carbonic acid, is a colorless gas with a boiling point of 8 °C. It is highly toxic, with a lethal concentration (LC_{50}) of 500 ppm over one minute or 5 ppm over one hour. The industry had to take special precautions to handle phosgene safely. One measure is to minimize phosgene at the isocyanate production plant. Phosgene, produced from CO and Cl_2, is fed into the phosgenation reactor immediately after production.

Cold-hot-phosgenation

The "cold-hot-phosgenation" or "base-phosgenation" is the most important phosgenation process used in industry. First, the amine and phosgene are dissolved in an inert solvent, generally o-dichlorobenzene. Subsequently, the amine solution is added to the phosgene solution under vigorous stirring at temperatures below 60 °C ("cold phosgenation"). The amine reacts with phosgene to carbamoyl chloride, liberating HCl, which reacts instantaneously with free amine to form the corresponding ammonium chloride. To prevent the undesired reaction of the amine with the already formed isocyanate or carbamoyl chloride to form urea, the amine is dispersed as quickly as possible in an excess of phosgene solution using specialized mixing devices. The temperature is subsequently raised to 180 °C ("hot phosgenation") to decompose carbamoyl chloride into isocyanate and HCl. Under these conditions, the less reactive ammonium chloride also undergoes phosgenation to form the corresponding isocyanate. After the reaction is complete, the solvent, excess phosgene, and HCl (two moles per mole of isocyanate) are stripped off and separated by fractional distillation/condensation. The solvent and phosgene are recirculated in the phosgenation process, and HCl is reused to produce chlorine.

Gas phase phosgenation

The conventional phosgenation process in solution has some significant disadvantages. It requires a substantial amount of solvent to run, and solvent evaporation after the reaction consumes a significant amount of energy. These disadvantages were overcome using the gas phase phosgenation process.

In the gas-phase reaction, the reactants are heated to approximately 300 °C and injected into a cylindrical reactor, where small amounts of solvent are added to dilute the reactive components and quench the gaseous reaction mixture. The low-boiling components, mainly HCl and excess phosgene, are discharged via the top of the reactor. The isocyanate and solvent are collected at the bottom and ready for further purification. Compared to conventional phosgenation, it saves about 80% of solvent and 60% of energy. Gas-phase phosgenation can be used to produce low-boiling isocyanates, such as aliphatic diisocyanates and TDI.

2.1.1 Methylene diphenyl diisocyanate

The precursor of methylene diphenyl diisocyanate (MDI) is aniline, which is produced from benzene. The first step in producing aniline is the reaction of benzene with a mixture of concentrated nitric acid and concentrated sulfuric acid, resulting in the formation of nitrobenzene. Subsequently, nitrobenzene is hydrogenated in the presence of a Raney nickel catalyst to form aniline.

The condensation reaction of aniline with formaldehyde, catalyzed by hydrochloric acid, forms a mixture of the diamine and polyamine oligomers. E.C. Wagner (1954) was the first to describe the complex mechanism of the condensation reaction, which consists of three main steps: condensation, the first rearrangement, and the second rearrangement.

During the pre-condensation step, aniline and formaldehyde react to form the unstable N-hydroxymethyl aniline. This intermediate condenses with a second aniline molecule, forming the dianilinomethane and water as a by-product.

The first rearrangement step begins with the protonation of the dianilinomethane, which reversibly dissociates into free aniline and the carbocation (1).

The carbocation (1) reacts with aniline under electrophilic substitution of aromatic hydrogen, which occurs at the electron-rich para and ortho positions, leading to the two isomeric aminobenzylanilines, namely p-ABA and o-ABA. Because of the steric hindrance at the ortho position, the formation of the para-substituted species is favored. As soon as the first molecules of ABA have formed, they compete against aniline as a reactant for the carbocation (1). Thus, higher oligomers can be formed depending on the excess of aniline.

The mechanisms of the first and second rearrangement are similar. In the second rearrangement, p-ABA and o-ABA are protonated to form the carbocations (2a) and (2b), respectively, which react with free aniline to yield the three methylene di-aniline (MDA) isomers; the 2-ring isomers 4,4'-, 2,4'-, and 2,2'-MDA. The carbocations can also react with these two-ring MDA molecules, forming higher homologs with an increased number of aromatic rings per molecule. Because the carbocations (2a) and (2b) are more stable and thus less reactive than the carbocation (1), the second rearrangement step is usually performed at higher temperatures. Although the reaction at the para position is favored over the ortho position, the high temperatures needed to run the reaction counteract the selectivity. As a result, increased temperatures yield increased amounts of ortho-isomers.

The aniline-to-formaldehyde ratio is the most critical parameter for controlling the oligomer mass distribution. Increasing the aniline-to-formaldehyde ratio reduces the formation of oligomers. At the same time, the ratio of 2,4'-MDA to 4,4'-MDA increases. Tab. 2.1 presents the crude MDA compositions for typical aniline-to-formaldehyde ratios ranging from 2.00 to 2.85 [1].

After neutralizing the reaction and evaporating the excess aniline, the MDA product is phosgenated to form the corresponding crude MDI. The oligomeric footprint of the MDA is found back in the crude MDI. The weight percentage of 2-ring MDI in crude MDI

Tab. 2.1: Composition of crude MDA.

Aniline excess (molar)	Monomeric MMDA (wt.%)	Oligomeric PMDA (wt.%)	Monomer composition (wt.%)	
			4,4'-MDA	2,4'-MDA
2.00	68	32	98.0	2.0
2.50	75	25	97.0	3.0
2.85	79	21	95.3	4.7

is approximately 50 wt.%, and the higher homologs follow a descending order with an increasing number of aromatic rings in the oligomer [Tab. 2.2]. The ratio of the 3-ring to 4-ring, 4-ring to 5-ring, and so on, is approximately constant; in the present example, it is approximately 2.

Tab. 2.2: A typical composition of crude MDI.

MDI type	Content (wt.%)
Monomer MDI (2-ring)	50
Polymer MDI (3-ring)	25
Polymer MDI (4-ring)	13
Polymer MDI (5-ring)	6
Polymer MDI (>5-ring)	6

A certain amount of monomeric MDI is removed from the crude MDI, for example, by thin film evaporation, leaving a mixture enriched in oligomeric polyisocyanates as the bottom product. This product is referred to as polymeric MDI, or short PMDI. The distillation is run until the bottom product has reached a specified viscosity. For most producers, their standard PMDI has a viscosity of 200 mPa·s (25 °C) and an average functionality of approximately 2.7.

The composition of PMDI can be determined by combining Gas Chromatography (GC) and Gel Permeation Chromatography (GPC). Using GC, the isomer ratio of the three 2-ring MDI isomers and four of the seven possible isomers of the 3-ring MDI can be quantified [Fig. 2.1]. Concerning the three 2-ring isomers, 4,4'-MDI exceeds 2,4'-MDI, and 2,2'-MDI is only present in small amounts. A closer look at the 3-ring isomers shows that one is dominant (tri-iso-1), whereas the other isomers are formed in small amounts.

The GC method cannot detect the higher oligomers because their boiling points are too high. GPC can detect the higher homologs but cannot separate the 2- and 3-ring isomers. Using GPC, compounds with up to 4 rings can be detected. Oligomers with five or more rings appear as a non-resolved broad peak [Fig. 2.2].

16 — 2 Starting components

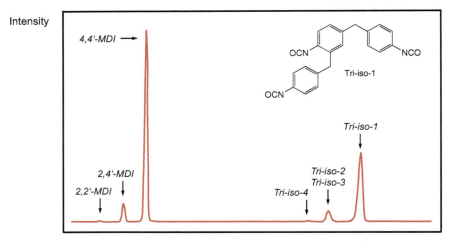

Fig. 2.1: GC of polymeric MDI, showing monomeric MDI and 3-ring isomers (with the kind permission of Covestro AG).

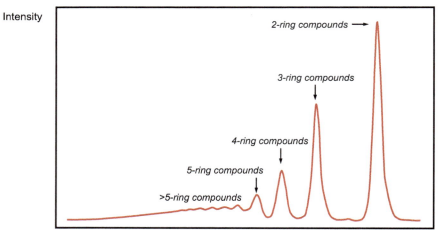

Fig. 2.2: GPC of polymeric MDI (with the kind permission of Covestro AG).

Monomeric MDI

The monomeric MDI stream obtained from the distillation process of crude MDI is further purified by distillation, yielding 4,4'-MDI, also known as pure MDI, as well as mixtures of 2,4'-MDI and 4,4'-MDI. The composition of the mixed isomer products depends on the producer, varying from 30:70 to 60:40, respectively. Pure MDI is produced with a purity of greater than 98% and is a crystalline product with a melting point of 39.5 °C. The mixed isomers, however, are liquids at room temperature (melting point 15 °C). Both products, pure MDI and the isomer mixtures, are commercially available. The former is predominantly used in elastomer applications, while the latter is used in flexible foam, coating, and adhesive applications.

The solid nature of 4,4'-MDI can be a disadvantage, as liquid isocyanates are preferred for processing. Two methods are used to liquefy 4,4'-MDI: pre-polymerization and uretonimine modification. Liquid 4,4'-MDI products are obtained when about 5 – 10% of the NCO groups are catalytically converted to carbodiimide. The carbodiimide, in turn, reacts readily with the isocyanate group of a third MDI molecule to uretonimine. The formed uretonimine molecule, with its bulky structure, is a potent liquefier, and its presence turns the uretonimine-modified isocyanate into a liquid product at room temperature.

Commercially available uretonimine-modified products typically have NCO contents of 29.5 wt.% (compared to 33.6 wt.% for 4,4'-MDI) and exhibit melting points between 10 and 15 °C. The oligomer composition, estimated using Monte Carlo simulations, consists of approximately 75 wt.% 4,4'-MDI, 18 – 20 wt.% uretonimine, and 5 – 7 wt.% higher oligomers. The average functionality of the isocyanate is about 2.2.

Another method of liquefying 4,4'-MDI is pre-polymerization with low-molar-mass diols. These prepolymers typically have NCO values of about 23 wt.%, and most suppliers offer such prepolymers. However, the most common prepolymers of MDI are based on high-molar-mass polyols and are used in various applications, such as flexible foams, elastomers, sealants, and adhesives.

Polymeric MDI (PMDI)

The viscosity of the isocyanate increases with the increasing content of higher molar mass oligomers, and the various grades of PMDI are characterized by their respective viscosities. The producers adjust the isocyanate viscosities (at 25 °C) to 100, 200, 400, 500, or 700 mPa·s. The isocyanate with a viscosity of 200 mPa·s is the standard polymeric grade of most producers. The PMDIs typically contain 30 – 70 wt.% monomeric (2-ring) MDI, 15 – 40 wt.% 3-ring MDI, and 15 – 30 wt.% higher oligomers. With the increasing viscosity of the isocyanate, the NCO value decreases from approximately 32 to 31 wt.%, while the average functionality of the isocyanates increases from 2.3 to 2.9. PMDI is predominantly used in the production of rigid foam and wood binders.

2.1.2 Toluene diisocyanate

Toluene diisocyanate (TDI) is produced in a three-step synthesis starting from toluene. First, toluene is treated with a mixture of concentrated nitric acid and concentrated sulfuric acid to form dinitrotoluene.

The reaction product consists of two isomers, 2,4- and 2,6-dinitrotoluene, in a mass ratio of 80:20, and up to 4 wt.% of the two vicinal dinitro compounds.

Next, the dinitrotoluene product is hydrogenated to form the corresponding toluene diamine (TDA). Before phosgenation, the two vicinal diamines must be removed as they would form cyclic ureas, which are solids and insoluble in TDI. The vicinal TDAs can be used as amine starters for producing polyether polyols.

Finally, the purified TDA is phosgenated, giving the desired diisocyanate. However, small amounts of solid by-products, mainly biurets, are formed. These by-products are removed during distillation and discharged. Based on toluene, the total yield of the isocyanate is 94%. TDI is a colorless liquid consisting of a mixture of 2,4- and 2,6-isomers in a ratio of 80:20, with a boiling point of 121 °C at 1.33 kPa.

The 80:20 mixture of the 2,4- and 2,6-isomers (T80) is the standard TDI used in industry. The NCO groups in the ortho-positions (2 and 6) are less reactive than those in the para-position (4). Because some applications may require faster- or slower-reacting TDI products, standard TDI is fractionated into two product streams: pure 2,4-TDI (T100) and a 65:35 mixture of 2,4- and 2,6-TDI (T65), representing the faster- and slower-reacting isocyanate, respectively [Tab. 2.3]. TDI is predominantly used in flexible foam applications. TDI prepolymers are used to produce soft elastomers and adhesives.

Tab. 2.3: Commercially available TDI isomer mixtures.

Isomer mixture	Abbreviation	mp (°C)
Pure 2,4-TDI	T100	21.8
2,4-TDI:2,6-TDI (80:20)	T80	13.5
2,4-TDI:2,6-TDI (65:35)	T65	5

2.1.3 Naphthalene-1,5-diisocyanate

The diisocyanate 1,5-naphthalene diisocyanate (NDI) is industrially manufactured in a multi-step synthesis starting from naphthalene. The diamine is produced via two different routes: the Bucherer reaction and direct amination. The reaction scheme of the Bucherer reaction is given below.

The reaction of naphthalene with fuming sulfuric acid at temperatures below 25 °C yields naphthalene-1,5-disulfonic acid, a strong acid known as "Armstrong's acid," named after Henry Edward Armstrong (1848–1937). The fusion of Armstrong's acid in sodium hydroxide results in the formation of the disodium salt of 1,5-dihydroxynaphthalene, which is transferred to the free diol upon acidification. The diol is converted in the presence of ammonia and sodium bisulfite into 1,5-diaminonaphthalene (Bucherer reaction), which can be phosgenated to NDI in a subsequent reaction.

The diamine can also be obtained by direct nitration of naphthalene, which yields an approximately 1:2 mixture of two isomeric products: 1,5-dinitronaphthalene and 1,8-dinitronaphthalene. The former is used for manufacturing 1,5-NDI, and the latter is a precursor to a red dye. NDI is utilized in the manufacture of high-quality elastomers. It is first prepolymerized and then chain-extended to produce the polymer.

2.1.4 NCO-prepolymers

The reaction of a polyol with excess isocyanate yields isocyanate prepolymers. Two-functional isocyanates are generally used; however, some applications may require PMDI-based prepolymers. The polyols are generally diols, but they can also be three-functional polyols. The hydroxyl group of the polyol reacts with the NCO group of the isocyanate, forming a urethane bond and terminating the polyol with an isocyanate group. With increasing excess of isocyanate, the amount of free isocyanate increases, reducing the prepolymer's viscosity. The polyol can, in principle, be taken from the system's polyol component, which is then reacted with the isocyanate. One of the main reasons for pre-polymerization is the liquefication of the isocyanate. This is especially relevant for solid isocyanates such as 4,4'-MDI and NDI. Furthermore, prepolymers are utilized to address processing challenges when manufacturing flexible foams and elastomers.

The basic process of producing PU, known as the one-shot process, involves the direct mixing of polyol and isocyanate. The polyol component in flexible foam and elastomer applications is typically the more significant volume stream and the higher-viscosity component. Moreover, it may be poorly compatible with isocyanate. As a result, mixing the components may be insufficient, leading to the formation of a poor-quality PU polymer. The use of prepolymers can improve the mixing efficiency. Pre-polymerization increases the polarity and viscosity of the isocyanate, and provides a closer match of the volume streams of polyol and isocyanate. Furthermore, using prepolymers reduces the reaction exotherm and facilitates the polymerization process. The rate of gel build-up is increased, and the phase separation process is delayed.

The viscosity of prepolymers increases over time due to side reactions. Consequently, the storage stability of isocyanate prepolymers is an issue. It depends on the type of prepolymer, the prepolymerization process, and storage conditions. The prepolymer stability improves with increasing polarity of the polyol and decreasing amounts of free isocyanate. The polyols must be dry, and catalysts are usually not applied in the prepolymerization reaction. After the prepolymerization reaction, acid stabilizers such as acid chlorides can be added to suppress side reactions, further increasing storage stability. Most standard MDI-based prepolymers have a shelf life of approximately six months. NDI prepolymers, however, which must be stored at elevated temperatures to avoid crystallization, are only stable for a few hours.

An important characteristic of isocyanate prepolymers is their molecular mass distribution. When reacting an excess of diisocyanate with a diol, the reaction mixture predominantly contains 2:1 adducts formed by the reaction of the two OH groups of one glycol molecule with two diisocyanate molecules. However, it also contains higher oligomers and cycles because the NCO groups of the adduct compete with the free diisocyanate molecules for reaction with the remaining hydroxyl groups. Chain extension and cycle formation are dependent on the isocyanate-to-polyol ratio. The oligomer distribution follows a Schulz-Flory distribution and can be simulated using statistical chain growth programs, such as Monte Carlo simulations.

The calculations are exemplified for a system based on 4,4'-MDI and PPG-2000, in which the molar ratio of isocyanate to polyol was increased stepwise from 2:1 to 20:1.

The free MDI content and oligomer composition of the prepolymers are given in weight percent (wt.%). [Tab. 2.4]. Considering the 2:1 prepolymer, in principle, two MDI molecules are available to react with the hydroxyl groups at both ends of a polyol molecule. The calculations indicate that the prepolymer is strongly chain-extended, and that the 2:1 adduct accounts for only 27 wt.% of the total prepolymer composition. Furthermore, 5 wt.% of free MDI is present, which is a direct consequence of the chain extension reaction. Chain extension requires less MDI per polyol molecule, leaving some free MDI. With an increasing ratio of MDI to polyol, chain extension and cycle formation are reduced. At the same time, the relative amount of the 2:1 adduct and free MDI in the oligomer mixture is increased. The reduction in higher oligomers and the increase in free MDI result in a decrease in prepolymer viscosity.

Tab. 2.4: Simulated compositions of 4,4'-MDI/PPG-2000 prepolymers at various NCO:OH ratios.

Prepolymer raw materials			Prepolymer composition (wt.%)						
Isocyanate	Polyol	Ratio (molar)	NCO content	MDI Monomer	Oligomers				Cycles
					2:1	3:2	4:3	higher	
4,4'-MDI	PPG-2000	2:1	3.4	5	27	23	16	23	6
		5:1	10.3	25	51	16	5	1	2
		10:1	16.8	45	46	7	1	≈ 0	1
		20:1	22.8	64	32	3	<1	≈ 0	<1

2.1.5 NCO-value and equivalent mass

Isocyanates are characterized by their NCO value which indicates the weight percentage of NCO groups in the isocyanate. The NCO value is experimentally determined by reacting the isocyanate with excess di-n-butylamine to form urea. The excess amine is back-titrated with hydrochloric acid (ISO 14896). The NCO value, expressed as wt.% NCO, can be calculated from the amount of amine that has reacted with the isocyanate. The NCO group has a molecular mass of 42 g/mol.

The NCO value can be used to calculate the isocyanate equivalent mass (EM), which represents the portion of the molar mass that corresponds to one isocyanate group. The molecular mass (M) of the isocyanate can be calculated, provided its functionality (f_n), which is defined as the number of NCO groups per molecule, is known.

$$EM = \frac{42 \cdot 100}{NCO - value} \qquad M = \frac{42 \cdot 100 \cdot f_n}{NCO - value}$$

For isocyanates with known molecular structure, the theoretical NCO value can be calculated from its molecular mass and functionality:

$$NCO\ value = \frac{42 \cdot 100 \cdot f_n}{M}$$

For example, 4,4′-MDI has a molecular mass of 250 g/mol and a functionality of 2. This yields an NCO value of 33.6 wt.% and an EM of 125 g/mol.

Most industrially used polyisocyanates have a functionality of two [Tab. 2.5]. Polymeric MDI (PMDI) is an exception, with an average functionality of 2.3 – 2.9.

Tab. 2.5: Key values for common polyisocyanates.

Polyisocyanate	Molar mass (g/mol)	Functionality	NCO value (wt.%)	Viscosity* (mPa·s)	mp (°C)
2,4-TDI	174	2	48.3	≈ 3	22
4,4′-MDI	250	2	33.6		39.5
4,4′-MDI:2,4′-MDI (50:50)	250	2	33,6	12	15
PMDI	290–360	2.3–2.9	30–32	200–700	–
NDI	210	2	40.0		127
TODI	264	2	31.8		71
HDI	168	2	50.0	≈ 3	−67
IPDI	222	2	37.8	≈ 15**	−60
H$_{12}$MDI	262	2	32.1	30	≈ 2–5
TMXDI	244	2	34.4		−10

*) at 25 °C; **) at 20 °C.

2.2 Polyether polyols

A wide range of polyols is available to produce polyurethanes. Their chemical composition, functionality, and molar mass largely determine the properties of the resulting polymers. Polyether polyols are cost-effective and versatile and can be applied in all major PU applications. Although polyurethanes were initially almost exclusively produced using polyester polyols, polyether polyols have now become the dominant choice.

Polyether polyols are manufactured through the reaction of a low-molecular-mass polyol or polyamine, referred to as the starter or initiator, with alkylene oxides, specifically ethylene oxide (EO) and propylene oxide (PO). The polyols are telechelic, meaning that the alcohol groups are positioned at the chain end. The number of hydroxyl groups on the starter and the produced polyol is equal, which gives the polyol functionality (f_n). The functionality of an amine-based polyol is twice the number of amine groups on the starter because each amine group can initiate two polyether chains. Polyether polyols with different molecular architectures can be produced. The molar mass and functionality can be altered by using different starters and varying the amounts of alkylene oxide per starter functional group. Furthermore, the alkylene oxides EO and PO can be copolymerized in several ways. The molar mass of the polyol is determined by the molecular mass of the starter and the mass of alkylene oxide monomers reacted onto the starter. The polyol equivalent mass (*EM*), the portion of the molar mass containing one hydroxyl group, is then the molar mass of the polyol (*M*) divided by its functionality (f_n):

$$EM = \frac{M}{f_n}$$

The functionality of the starter ranges from 2 to 8. Two-functional polyols can be produced from the following starters: ethylene glycol, diethylene glycol, 1,2-propylene glycol, dipropylene glycol, and bisphenol A. Three-functional polyols can be prepared from trimethylolpropane and glycerine, and four-functional polyols can be synthesized from pentaerythritol, ethylenediamine, and o-toluenediamine. The use of sorbitol and sucrose gives six and eight-functional polyols, respectively. The most common starters are depicted below.

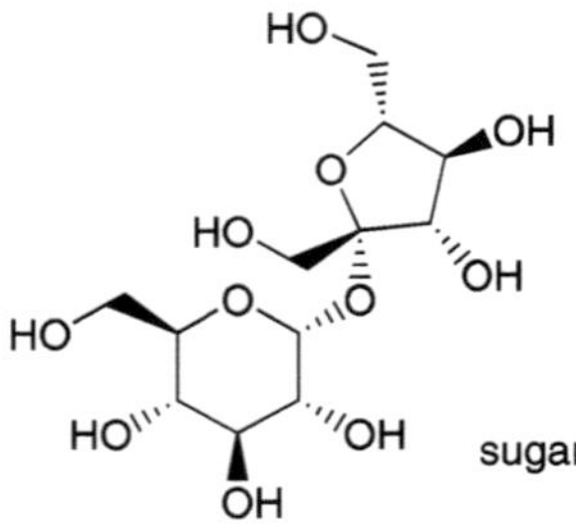

sugar

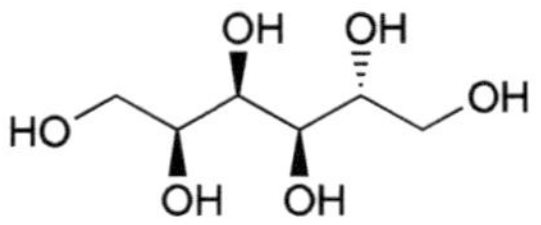

sorbitol

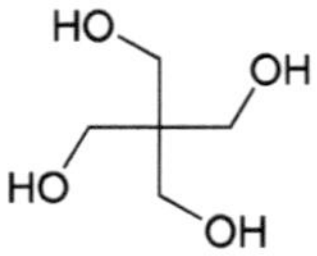

trimethylol propane

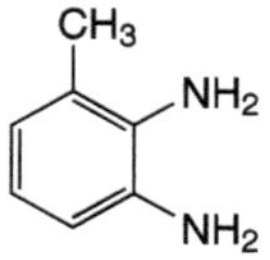

o-toluylenediamine

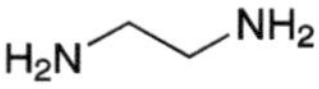

ethylenediamine

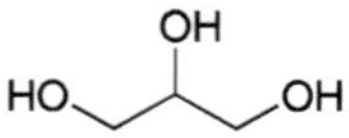

glycerine

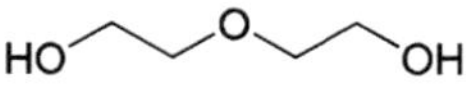

diethylene glycol

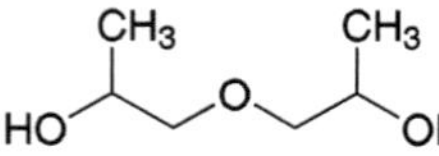

dipropylene glycol

The viscosity of the polyol increases with the increasing functionality of the starter and decreases with the number of alkylene oxide molecules grafted onto the starter. Polyols with high functionality and a low number of alkylene oxide groups per hydroxyl functional group will exhibit high viscosities. Such polyols are commercially relevant and used in rigid foam applications. The viscosities of polyols containing high-functionality starters can be reduced by combining high- with low-functionality starters, such as sucrose with glycerine or diethylene glycol.

2.2 Polyether polyols — 25

2.2.1 Base-catalyzed polyether production

The most common manufacturing process for polyether polyols involves the anionic polymerization of an alkylene oxide, initiated by a starter molecule.

The nucleophilicity of the hydroxyl group is insufficient to allow a direct reaction with the alkylene oxide. Its nucleophilicity can be increased by forming an alcoholate, generally accomplished by adding potassium hydroxide and removing water.

The oxirane has substantial ring strain, making epoxides highly reactive. The alcoholate attacks one of the two partially positively charged carbon atoms of the oxirane ring, followed by ring opening. When using EO, primary OH groups are formed. Using PO, the alcoholate group attacks the carbon atom with the lowest steric hindrance, forming a secondary hydroxyl group. PO shows lower reactivity in the alkoxylation reaction than EO due to steric hindrance and the positive inductive effect of the methyl group.

Amine-containing compounds can also serve as starting materials. The addition reaction proceeds without a catalyst because the NH-group is a sufficiently strong nucleophile to initiate the ring-opening reaction. An inner salt (zwitterion) is formed as an intermediate that rearranges to the corresponding hydroxyalkyl amine by proton transfer from the nitrogen to the oxygen atom. The secondary NH group is more nucleophilic than the OH group, so adding a second alkylene oxide occurs at the nitrogen atom, forming an alkyl-substituted dialkanolamine. The dialkanolamine can react with additional alkylene oxide; however, this only occurs after "restarting" the reaction by adding a base such as KOH.

The polyols derived from aminic starters contain tertiary amine groups that can catalyze the PU reaction. Thus, polyols based on aminic starters are more reactive in the reaction with isocyanate than those based on hydroxylic starters.

The potassium hydroxide is added to the starter as an aqueous solution. Water, however, can react with alkylene oxide to form glycol. Further reaction of the glycol with alkylene oxide yields a two-functional polyol – a diol. Therefore, water must be removed from the starter composition before starting the alkoxylation reaction to avoid diol formation. In practice, traces of water will always be present in the starter mixture, and the final polyol will contain some diol.

Under base conditions, propylene oxide can undergo a side reaction, forming terminal unsaturated mono-functional alcohols. The alkoxide can abstract a proton from the methyl group of propylene oxide, and after rearrangement, allyl alkoxide is formed. The allyl alkoxide, in turn, can initiate the polymerization of a new polymer chain. The polyol formed has a hydroxyl functionality of one and is called mono-ol. Its presence interferes with the polymer network build-up in the PU reaction and introduces dangling ends in the polymer network.

The presence of monofunctional alcohols is insignificant in low-molar-mass polyols but becomes increasingly substantial with increasing molar mass. The molecular composition of the polyol is determined by the ratio of the polymerization rate constant and the allyl alkoxide formation rate constant. This ratio depends on the reaction temperature and the type of hydroxide used. Mono-ol formation decreases with reduced reaction temperatures and in the series of hydroxides based on $Na^+ > K^+ > Cs^+$. Mono-ol formation limits in practice the molar mass of propylene glycols. The molar mass is limited to approximately 2,000 g/mol using potassium hydroxide as a catalyst.

Consider an all-PO glycerine-based polyol with a molecular mass of 6,000 g/mol. Its theoretical functionality is three. In practice, however, the polyol is a complex mixture of triols, diols, and mono-ols. The exact composition depends on the reaction conditions applied and the catalyst used. Typically, the polyol consists of approximately 90 wt.% triol with about 5 wt.% of diol and mono-ol each. The presence of the side products reduces the average functionality of the polyol. In analogy to the number average molar mass (M_n), the number average functionality (f_n) is defined as:

$$f_n = \frac{\sum n_i \cdot f_i}{\sum n_i}$$

where n_i is the number of molecules with functionality f_i. The average functionality of the example glycerine polyol amounts to about 2.5.

Industrially, polymerization reactions are run as discontinuous batch processes in reactors with reaction volumes ranging from 10 to 100 m^3. First, the starter is mixed with an aqueous potassium hydroxide solution under a nitrogen atmosphere. The alcoholate is formed after the removal of water. Subsequently, PO and/or EO are added continuously under external cooling, while maintaining the reaction at approximately 3 – 5 bar and between 90 and 130 °C. After the reaction is completed, the catalyst must be neutralized. This can be achieved by adding an acid or an acid ion-exchange resin. Using sulfuric and phosphoric acid, or carbon dioxide, has the advantage that the resulting potassium salts are insoluble in the polyether polyol and can be easily removed by filtration. Carboxylic acids, such as lactic acid and acetic acid, are also used for neutralization. The corresponding potassium salts are soluble in the polyol and not removed. These salts, however, show catalytic activity in the PU reaction. The acid-neutralized polyols are, therefore, not suitable for all PU applications; they are only used in those applications where increased PU reaction rates are acceptable. Finally, the product is stripped to remove volatile components. Antioxidants may be added to enhance the storage life of the polyol.

2.2.2 DMC-catalyzed polyether production

Double Metal Cyanide complexes (DMC) were discovered in the 1960s as effective catalysts for the propoxylation of alcohols, allowing the production of polypropylene polyols with low mono-ol contents and narrow molar mass distributions. They were further improved by developing the so-called IMPACT catalysts, which exhibit significantly higher reaction rates and yield polyols with even lower mono-ol contents. Due to their high reactivity, the necessary amounts of catalyst (in the low ppm range) are sufficiently low to remain in the final polyether polyol, making the process highly economical. Furthermore, the technology enables the manufacture of high-molar-mass polyols in the 20,000 g/mol range. The DMC technology is currently used to produce PO homopolymers for elastomer and adhesive applications, and slabstock polyols (triol polyols with equivalent masses of approximately 1,000 g/mol containing some randomly distributed EO units) for the flexible foam market.

DMC catalysts may be prepared by mixing an aqueous solution of zinc dichloride (or other zinc salts, such as zinc acetate) with an aqueous solution of a metal cyanide salt, e.g., potassium hexacyanocobaltate(III), to form the insoluble zinc hexacobaltate(III) salt.

$$Zn_3\left[Co(CN)_6\right]_2 \cdot x\,ZnCl_2 \cdot y\,H_2O \cdot z\,Ligand$$

The salt is usually complexed with low-molecular-mass ligands, such as glyme or alcohols (in particular tertiary butanol) during the purification step, or synthesis. A considerable excess of Zn^{2+} cations is applied to the hexacyanocobaltate to ensure that small crystals are formed with an excess of, for example, zinc dichloride at the surface.

DMC is a solid heterogeneous catalyst, with the reaction occurring on the material's surface. The propoxylation occurs in a sequence wherein the PO molecule coordinates to the catalyst surface, where its ring is opened by a nucleophilic attack of the hydroxyl group of a polyol chain [2]. Subsequent protonation, achieved through a proton shift, liberates the propoxylated alcohol, allowing catalysis to proceed from the start after PO coordination at the free site.

Low-molar-mass hydroxyl molecules can act as ligands to form stable complexes with the DMC catalyst, thus poisoning it for PO coordination. Therefore, conventional low-molecular-mass starters, such as glycerine, cannot initiate the DMC process. This can only be achieved with short-chain polyols above a critical molar mass, leading to a lower hydroxyl concentration. Indeed, the complexation strength of polyol chains on the catalyst surface decreases with increasing molar mass, most likely due to the lower hydroxyl content.

It was observed that short-chain polyols are propoxylated more rapidly than long-chain polyol molecules due to their higher hydroxyl content. A nucleophilic attack of a short-chain polyol with a higher hydroxyl content is more likely than that of a longer-chain polyol with a lower hydroxyl content. This causes the DMC catalysts to exhibit a special kinetic regime known as "catch-up kinetics". The kinetic effect is so predominant that polymerizing a mixture of polyols with different molar masses yields a high-molar-mass polyol with a narrow, almost monomodal molar mass distribution – as long as the chains are sufficiently short. This finding has been utilized to develop highly efficient production processes, where a short-chain polyol is continuously added as a starter to the reaction mixture.

2.2.3 Polyols containing EO and PO

The polyether polyols used in PU applications are predominantly PO-based. However, some EO can be copolymerized into the polyol structure to increase its reactivity and hydrophilicity. The following structures are synthesized:
– Homopolymers of all PO or EO
– Block-copolymer structures PO-EO
– Random co-polymers PO/EO

The short-chain polyether polyols for rigid foam applications are predominantly PO-based homopolymers. The long-chain polyols for flexible foam applications are either random PO/EO or block-copolymer polyols with a PO-based core and an EO cap.

The oxygen atom in the ether group of polyethylene oxide can interact with polar molecules, such as water, through hydrogen bonding. This interaction is limited in polypropylene oxide because of steric hindrance by the pendant methyl group adjacent to the ether oxygen atom. Incorporating EO monomers, therefore, increases the polyol polarity and hydrophilicity, aiding miscibility with other foam ingredients, such as water.

PO polyols exhibit secondary hydroxyl groups. These are about four times less reactive against isocyanates than primary hydroxyl groups. The reactivity of the PO-based polyol can be increased through capping or tipping with EO.

In the reaction with EO, the reactivity of the secondary hydroxyl group is lower than that of the primary hydroxyl group, which results in a non-selective reaction between the propylene glycol polyol and EO. This can be easily illustrated by the following example: a polyol with two secondary hydroxyl groups that reacts with two molecules of EO. The first molecule of EO has no choice but to react with one of the two secondary OH groups, forming a primary OH group. The second EO molecule now has the choice between the remaining original secondary OH group and the newly formed primary OH group. It will preferentially, but not exclusively, react with the primary OH. As a result, the primary hydroxyl content of the polyol does not progress linearly but asymptotically with the amount of EO used for tipping.

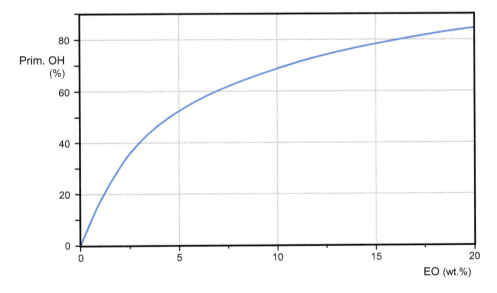

Fig. 2.3: Primary hydroxyl content versus EO content used for capping a trifunctional polyether (molecular mass of PO core polyol is 3,000 g/mol and K = 8).

The percentage of primary OH groups of an EO-tipped polyol can be calculated according to [3]:

$$\frac{[EO]}{OH_{total}} = (1-K)\cdot r - K \cdot \ln(1-r) \quad \text{with} \quad r = \frac{OH_{prim}}{OH_{total}}$$

In this equation, $[EO]$ represents the weight amount of EO in the final polyol, while (OH_{total}) and (OH_{prim}) represent the molar concentrations of all OH groups and the molar concentration of primary OH groups, respectively. The distribution factor (K) depends on the processing conditions. Fig. 2.3 illustrates the calculated primary OH content versus weight percentage of EO for an all PO-polyol with a molar mass of 3,000 g/mol and K = 8.

Triol polyols with molar masses between 5,000 and 6,000 g/mol, capped with ethylene oxide, are referred to as reactive or molded polyols and are used to produce flexible foam and elastomers. Their primary OH contents range from 70 to 85%, which can be achieved with EO capping using approximately 10 – 20 wt.% EO.

2.2.4 Filled polyether polyols

Polyether polyols containing solid polymer particles as the dispersed phase are primarily used in flexible foam applications to enhance hardness. Three different types of polymer-filled polyether polyols are available:
- Polymer polyol – styrene-acrylonitrile copolymer particles.
- PHD polyols – polyurea particles.
- PIPA polyols – polyurethane particles.

Polymer polyols

Styrene, acrylonitrile, a macromer, and a radical initiator are dissolved in a polyol. The polyol is a flexible foam polyol with molecular masses ranging from 3,000 to 6,000 g/mol, serving as the reaction medium. The macromer is prepared separately by reacting a vinyl monomer onto a polyol molecule. Initially, the system operates in a single phase. The radical initiator, e.g., azobis(isobutyronitril) (AIBN), copolymerizes the vinyl monomers. The growing SAN polymer precipitates when it reaches a critical molecular mass, forming tiny droplets. The copolymerized macromer anchors polyol molecules on the droplet surface and stabilizes the droplets through steric repulsion [Fig. 2.4]. With continuing reaction, the droplets grow and eventually solidify. The particles in the final dispersion show diameters ranging from 1 to 5 microns.

The viscosities (at 25 °C) of polymer polyols typically amount to 1,500 mPa·s at 25 wt.% filler loading or 5,000 mPa·s at 45 wt.%.

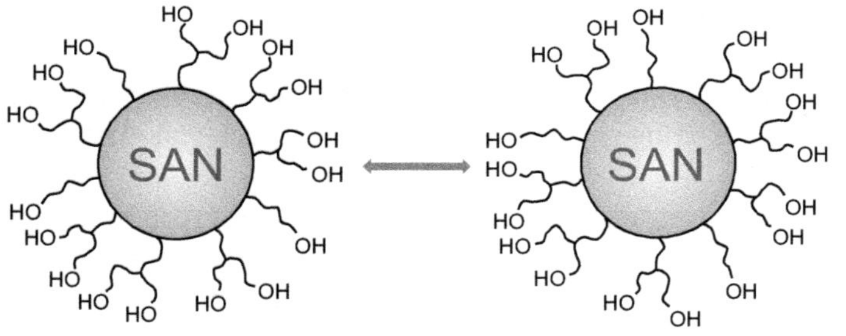

Fig. 2.4: Steric stabilization of SAN particles.

PHD polyols

PHD polyols (German: Poly-Harnstoff-Dispersion), also known as polyurea dispersions, contain poly(hydrazodicarbonamide) particles. They are produced by dissolving hydrazine in a polyether polyol and reacting it with TDI.

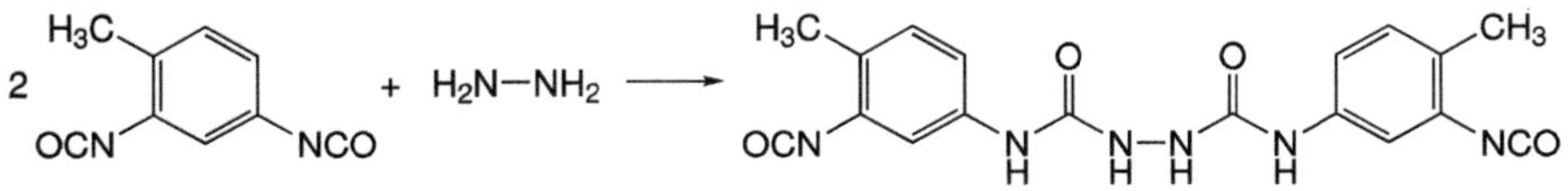

TDI reacts predominantly with hydrazine, forming poly(hydrazodicarbonamide), and to a minor extent, with polyol, which serves as the graft for particle stabilization. PHD dispersions show broader particle size distributions and higher viscosities than polymer polyols. For example, a filler loading of 30 wt.% typically yields viscosities of approximately 3,500 mPa·s (at 25 °C). Hence, the filler contents must be reduced to achieve comparable viscosities to polymer polyol. The use of PHD polyols in flexible foam reduces flammability.

PIPA polyols

In PIPA polyols, the filler is generated by reacting an alkanolamine with a polyisocyanate, forming a polyisocyanate polyaddition particle (Poly-Isocyanate Poly-Addition reaction; PIPA). Its production is often conducted at the PU manufacturer's production plant for captive use. The polyisocyanate, such as MDI or TDI, is mixed with an alkanolamine, typically triethanolamine, using a high-pressure mixing machine, and the resulting reaction mixture is then injected into a polyol. The dispersion can be used to manufacture PU foam straight after its production. The reaction and mixing conditions must be tightly controlled to produce phase-stable dispersions with acceptable viscosities.

2.2.5 Poly(tetramethylene ether) glycols

Poly(tetramethylene ether) glycol (PTMEG), also known as polytetrahydrofuran or poly(tetramethylene oxide), is primarily used to produce TPU, high-end elastomers, and "Spandex®" fibers.

PTMEG is produced by the ring-opening polymerization of tetrahydrofuran, catalyzed by Lewis acids, such as boron trifluoride, using water or a diol as the starter. It is commercially available with molar masses from 250 to 2,000 g/mol. PTMEG 1,000 and 2,000 are the preferred products for PU elastomer applications.

2.3 Polyester polyols

Polyester polyols for PU applications are produced from diacids and short-chain diols. The most important polyesters are based on aliphatic or aromatic dicarboxylic acids and aliphatic diols containing 2 – 6 carbon atoms or short glycols. Polyester polyols are used in technical and rigid foams, TPU, and cast elastomers.

2.3.1 Polyester polycondensation reaction

The polycondensation of dicarboxylic acids or their anhydrides with excess difunctional polyols yields polyester polyols. Because esterification is an equilibrium reaction, water must be continuously removed to shift the reaction toward the formation of the ester. In practice, only diols with primary hydroxyl groups are taken.

Common diols are ethylene glycol, diethylene glycol, 1,4-butanediol, 1,6-hexanediol, or neopentyl glycol. Small amounts of three- or four-functional monomers, such as trimethylolpropane and pentaerythritol, can be co-reacted to introduce branching. Introducing branching, however, is very limited because it strongly increases the polyol viscosity.

Aromatic and aliphatic diacids are used. The most commonly applied aliphatic diacid is adipic acid. Terephthalic acid, phthalic acid (generally used as the anhydride), and isophthalic acid are the most commonly applied aromatic diacids.

The ratio of hydroxyl to acid groups determines the molar mass of the resulting polyester polyol, which is only achieved after the complete conversion of the reaction.

High conversions are also needed to obtain polyester polyols with low acidity. Unreacted acid groups affect the reactivity of the PU reaction because they interfere with the catalysis. For instance, the commonly used tertiary amine catalysts form a salt with the acid, losing their catalytic activity.

Commercially produced polyester polyols always contain some unreacted acid groups. The acid value of the polyester polyol is obtained by direct titration with a KOH solution (ISO 2114). Analogously to the hydroxyl value, the acid value is expressed in [mg KOH/g] and gives the weight amount of KOH required to neutralize 1 g polyol. The acid value of polyester polyols must be below 3, preferably below 1 mg KOH/g.

The polycondensation reaction is carried out batch-wise in stirred tank reactors with volumes as large as 20 m^3. Because it is an equilibration reaction, it takes relatively long to reach the targeted conversion. Temperatures of approximately 150 – 250 °C and a high vacuum are required to complete the reaction in 24 – 40 h.

Like polyether polyols, polyester polyols can be categorized into two types: low-molar-mass polyesters with molar masses of up to 400 g/mol and high-molar-mass polyesters ranging from 1,000 to 3,000 g/mol. The low-molar-mass polyesters used for rigid foam applications are typically based on aromatic diacids to improve fire resistance. Furthermore, they may be slightly branched to improve the mechanical properties of the foams. The high-molar-mass polyesters used for technical flexible foams and elastomer applications are predominantly based on adipic acid and aliphatic diols containing 2 – 6 carbon atoms.

Polycondensation yields polyesters with broad molecular mass distributions (dispersity index equals 2). They contain considerable amounts of unreacted diol and can also contain non-functional cyclic polyester molecules. The cyclic components can be removed by thin film evaporation at low temperatures. However, over time, e.g., during storage, they can reform through transesterification.

Polyester polyols can be either crystalline or amorphous, depending on the structure of the monomers. A polyester polyol with a regular structure, such as one produced from 1,4-butanediol and adipic acid, is a crystalline solid at room temperature with a melting point of approximately 50 °C. The crystallization can be suppressed by using mixtures of glycols, reducing the regularity of the structure. Due to ester-ester interaction, the viscosity of polyester polyols is higher than that of polyether polyols with comparable molar masses.

2.3.2 Polycaprolactone polyols

Whereas the standard polyester polyols are characterized by "head-to-head" and "tail-to-tail" configurations, polycaprolactone is a head-to-tail polyester polyol. It is industrially produced by ring-opening polymerization of caprolactone onto an OH-functional starter in the presence of metal-organic catalysts, such as tin or titanium alkoxides.

Commercially available polycaprolactones for PU applications typically have a functionality of two and molar masses ranging from 500 to 2,000 g/mol. These polyesters have relatively narrow molecular mass distributions and are especially suitable for manufacturing high-performance elastomers.

2.3.3 Polycarbonate polyols

There are two routes to polycarbonate polyols: the reaction of diols with dimethyl carbonate and the reaction of epoxide with carbon dioxide. The former produces specialty polyols, while the latter is predominantly used to produce polyols for flexible foam applications.

The production process for carbonate specialty polyols uses the transesterification reaction of dimethyl carbonate with aliphatic diols, generally 1,6-hexanediol.

Polycarbonate polyols are mainly used to manufacture adhesives, artificial leather, or high-performance elastomers. Their molar masses are typically 1,000 and 2,000 g/mol. When hexanediol is used, the resulting polycarbonate polyols are waxy solids. Liquid polyols can be obtained using longer diols or diol mixtures.

An alternative route to polycarbonates is based on the ring-opening copolymerization of propylene oxide with carbon dioxide.

During the reaction, cyclic carbonate is formed as a by-product by the insertion of carbon dioxide into the epoxide ring. DMC catalysts were developed to suppress this side reaction. Theoretically, up to 50 mol% of carbon dioxide (corresponding to 43 wt.%) can be incorporated into the polyol chain, resulting in an alternating copolymer. Such polymers, however, are semi-solids and challenging to apply. Therefore, the carbon dioxide content is reduced to give polyols with manageable viscosities. Tab. 2.6 presents examples of experimental and commercially available polyols. The latter two are used in conventional slabstock foam production.

Tab. 2.6: Polycarbonate polyols based on propylene oxide and carbon dioxide.

Product name (Company)	Converge® 212–10 (Aramco)	Converge® CPX-2,502-56 (Aramco)	Cardyon® LC 05 (Covestro)
CO_2 content	40 wt.%	20 wt.%	14 wt.%
Functionality	2	2	3
OH-value	112 mg KOH/g	56 mg KOH/g	56 mg KOH/g
Molar mass	1,000 g/mol	2,000 g/mol	3,000 g/mol
Viscosity	10,000 mPa·s (75 °C)	32,000 mPa·s (25 °C) 600 mPa·s (75 °C)	10,500 mPa·s (20 °C)

2.4 Bio-based polyols

The increasing pressure to move away from fossil resources has triggered the development of PU starting materials from renewable sources. This involves substituting standard crude oil-based precursors with the corresponding renewable bio-based analog and using naturally occurring raw materials as alternatives to fossil precursors.

2.4.1 Renewable precursors

Recently, advanced biotechnology and catalysis have opened up various opportunities to utilize biomass, such as sugar, corn, or lignin, as feedstock for producing drop-in precursors for manufacturing polyols.

The short-chain diols 1,3-propanediol and 1,4-butanediol can be obtained by fermentation of corn and sugar. They are used to produce 100% bio-based poly(trimethylene ether) polyol and PTMEG, respectively. Aliphatic diacids, such as succinic acid can be produced through fermentation of carbohydrates, whereas azaleic and sebacic can be produced from castor oil. Together with bio-based aliphatic diols, it allows the manufacturing of 100% renewable polyester polyols. Furthermore, various bio-based raw materials, such as cardanol (derived from cashew nut shells), are utilized to produce polyols.

2.4.2 Natural oil polyols

Natural oils, such as soybean, palm, and rapeseed, are esters derived from glycerine and three fatty acids. The fatty acids may vary depending on the type of oil. Below are some examples of saturated and unsaturated fatty acids.

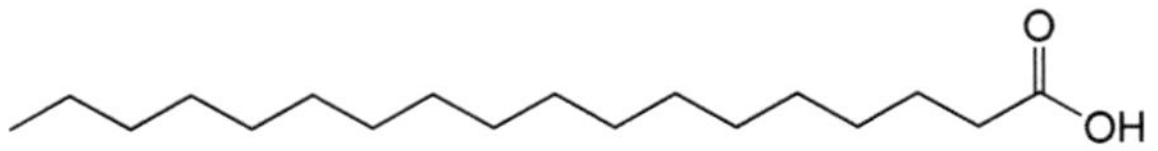

stearic acid
(C18:0)

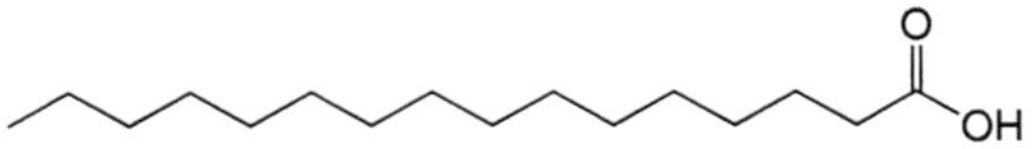

palmitic acid
(C16:0)

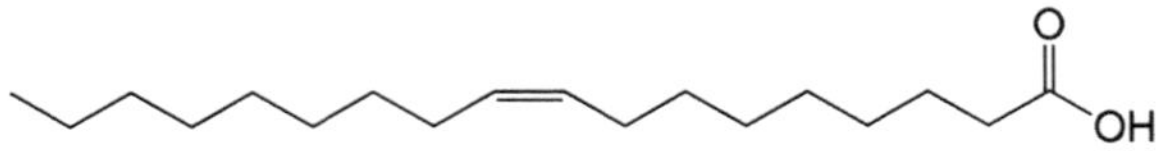

oleic acid
(C18:1; cis-9; omega-9)

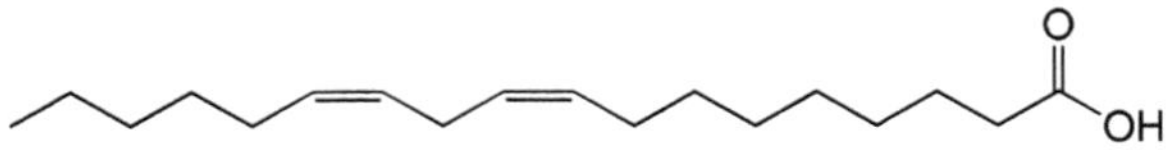

linoleic acid
(C18:2; cis-9,12; omega-6)

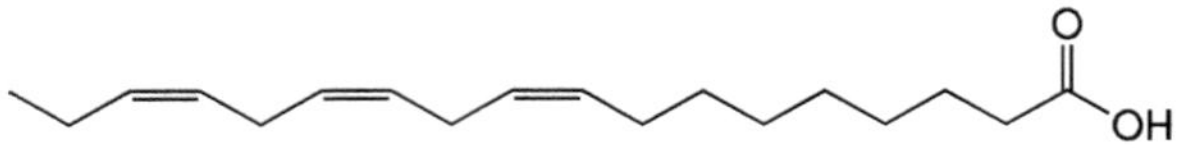

α-linolenic acid
(C18:3; cis-9,12,15; omega-3)

First, the oils are hydrolyzed, yielding glycerine and fatty acids. Glycerine can be used as a starter for polyether polyol production, and the fatty acids can be incorporated into polyester or hybrid polyester-ether polyols. The copolymerization of fatty acids increases the hydrophobicity and compatibility of the polyols with apolar foam ingredients, such as pentane.

Natural Oil-based Polyols (NOP) are produced by trans-esterification of triglycerides with polyols. Hydroxyl groups can be introduced into unsaturated fatty acids through the hydroformylation or epoxidation of the double bonds, followed by a subsequent ring-opening reaction with alkanol amines, water, or mono-alcohols. These reaction products can be modified further by subsequent alkoxylation. Dimerization of unsaturated fatty alcohols yields diols ("dimer diols"). Ozonolysis of unsaturated fatty acids yields mixtures of mono- and diacids. After separation, the diacids can be used as monomers to produce polyester polyols.

Castor oil is a natural polyol containing secondary hydroxyl groups. Its OH-value is 430 mg KOH/g, and its functionality is approximately 2.7. Castor oil can be used in rigid foams and coatings. It can be oxypropylated using DMC catalysis. These higher-molecular-mass polyols are used in flexible foam applications.

2.5 Hydroxyl value and equivalent mass

The structure of the polyol can be derived from the starting materials employed to produce it and the experimentally determined hydroxyl value, abbreviated as OH value.

The hydroxyl value is determined by acetylation of the polyol with excess acetic anhydride (ISO 4692–2). Each OH group of the polyol is esterified to the corresponding acetic acid ester, resulting in the formation of one free acetic acid molecule. After the reaction, the excess acetic anhydride is hydrolyzed with water, yielding two acetic acid molecules per anhydride. The acetic acid is titrated with a standard potassium hydroxide solution. The hydroxyl value is the number of milligrams of potassium hydroxide required to neutralize the acetic acid taken up on the acetylation of one gram of polyol, expressed as milligrams of KOH per gram (mg KOH/g). With the molecular mass of KOH (56.1 g/mol), the relationships between equivalent mass (EM), molar mass (M), and functionality (f_n) can be expressed as follows:

$$EM = \frac{M}{f_n} = \frac{56.1 \cdot 1{,}000}{OHv} \qquad M = \frac{f_n \cdot 56.1 \cdot 1{,}000}{OHv} \qquad OHv = \frac{f_n \cdot 56.1 \cdot 1{,}000}{M}$$

Fig. 2.5 outlines the various polyether and polyester polyol families, highlighting their functionalities and hydroxyl values. Crosslinkers, used to increase the crosslink density of PU polymers, are produced by adding 1 – 2 alkylene oxides per starter molecule, such as glycerine, and exhibit OH values between 700 and 1,000 mg KOH/g. Polyols for rigid PU foams have OH values of 300 – 600 mg KOH/g and average functionalities between 3 and 5. Flexible foams are produced from polyols with OH values ranging from 28 to 56 mg KOH/g and a nominal functionality of 3. Elastomers are produced from two-functional polyols with OH values ranging from 28 to 112 mg KOH/g.

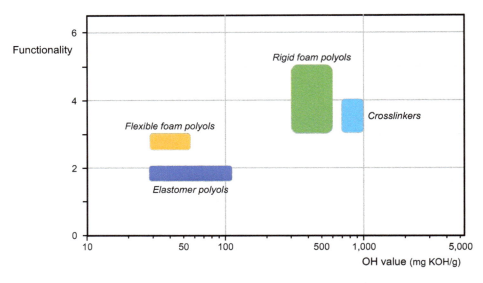

Fig. 2.5: Key values of polyether and polyester polyols used for the various PU applications.

2.6 Diamines and polyamines

The reaction of amines with isocyanate to form urea is much faster than the urethane reaction and does not need catalysis. Two groups of polyamines play a role in polyurethane chemistry: low-molecular aromatic diamines used as chain extenders and polyetheramine soft blocks.

2.6.1 Diamine chain extenders

The reactivity of unhindered aromatic diamines with isocyanates is too high for standard processing methods. Sterically hindered diamines and diamines with electron-withdrawing substituents react significantly slower and are used in PU applications.

Diethyl toluene diamine (DETDA) is the most commonly used sterically hindered aromatic diamine. It is a mixture of isomers, with the two main isomers listed below. It is based on the same diamine used to produce TDI, namely 2,4-/2,6-TDA in a ratio of 80:20. The diamine is reacted with 2 moles of ethene using strong Lewis acids as catalyst. DETDA-containing systems are slow enough to be processed using standard RIM machines.

The most common aromatic diamines with electron-withdrawing substituents are 2-methylpropyl 3,5-diamino-4-chlorobenzoate (below left) and 4,4′-methylene-bis(2-chloroaniline) (MBOCA) (below right). Recently, significant efforts have been made to replace MBOCA, as it is a Substance of Very High Concern (SVHC) and has been included in the European Chemical Agency (ECHA) candidate list for authorization.

2.6.2 Polyether amines

The hydroxyl groups of polyols can be converted into amino groups through reductive amination with hydrogen and ammonia under high pressure and elevated temperatures.

The amino groups of polyether amines react quickly with isocyanate, forming urea linkages. The high reactivity is an advantage in applications such as polyurea spray coatings. However, in polyurea RIM, the reactivity of the polyether amines has to be reduced. This can be achieved by blocking the amines with carbon dioxide to the corresponding carbamates. When reacting with isocyanates, carbon dioxide is liberated, acting as a blowing agent that assists in mold filling.

2.6.3 Mannich polyols

Aromatic Mannich polyols exhibit increased reactivity and are primarily used in PU spray foam applications. They are produced by alkoxylation, usually with PO, of aromatic Mannich bases. The Mannich bases, in turn, are formed through a three-component condensation reaction involving a phenol, such as p-nonylphenol, formaldehyde, and an alkanolamine, such as diethanolamine.

2.7 Additives

The blowing agents, surfactants, and fire retardants used in PU applications are described. Catalysts will be discussed separately in Chapter 3.2.7. Additives, such as antioxidants, fillers, pigments, colorants, release agents, and biocidal or fungicidal additives, are not discussed.

2.7.1 Blowing agents

Blowing agents are required to expand PU foams. The foams can be chemically or physically blown using chemical or physical blowing agents. Water is the most used chemical blowing agent. Its reaction with isocyanate yields carbon dioxide, which diffuses into the cells and expands the foam. Physical blowing agents are low-boiling inert liquids added to the polyol component or fed directly into the mixing head. The exothermic reactions evaporate the blowing agent, which expands the foam.

The first decades of industrial use of PU were dominated by blowing with water. The invention of physical blowing agents in the late 1950s marked the beginning of the success story of rigid foams, as their use significantly improved their thermal insulation value. Rigid foams are closed-cell foams that keep the blowing agent in the foam. Physical blowing agents show lower heat conductivity values than carbon dioxide and air. Consequently, the heat conductivity values of the foams are improved when physical blowing is used. In the early days, blowing agents were also used as processing agents for foam expansion, such as open-cell flexible foam, where the blowing agent escapes after the foam expansion is complete. Nowadays, for environmental reasons, physical blowing agents (except carbon dioxide in flexible foams) are only used in applications where they provide additional benefits, such as good heat insulation values in rigid foam.

The physical blowing agents should have a low boiling point, high molecular mass, and poor PU swelling characteristics. Ideally, the boiling temperatures should be between 20 and 40 °C, allowing evaporation and foam expansion early in the reaction. The molecular mass of the blowing agent should be high because the thermal conductivity of the gas decreases with increasing molecular mass. Furthermore, the blowing agent should be apolar to minimize swelling of the polar PU polymer. Swelling leads to the softening of the polymer matrix, thus deteriorating the mechanical properties.

The first employed physical blowing agents were chlorofluorocarbons (CFCs). The prime contender in this class of blowing agents was trichlorofluoromethane (CFC-11). It had all the physical properties required for a good blowing agent and possessed an exceptionally low thermal conductivity. Chlorofluorocarbons, however, exhibit a high Ozone-Depleting Potential (ODP). The pressure to move away from chlorofluorocarbons was accelerated after the ratification of the Montreal Protocol (in 1987), which proposed a phase-out plan for ozone-depleting substances. Worldwide, R&D programs were initiated to develop alternative blowing agents. However, several early alternatives exhibited a high Greenhouse Warming Potential (GWP). To better compare the ODP and GWP values of alternative blowing agents, a relative scale was introduced, with the ODP of CFC-11 and the GWP of carbon dioxide set to 1.

Blowing agents are characterized using the following nomenclature, where the free bonding positions at carbon are saturated with chlorine.

R-ABCD

R stands for refrigerant

A: number of double bonds

B: number of carbon atoms minus 1

C: number of hydrogen atoms plus 1

D: number of fluorine atoms

When isomers can be formed, the structure is determined by additional letters at the end, e.g., R-134a. The following examples illustrate the nomenclature.

CCl_3F is called R-11 because it contains no double bonds (A = 0), one carbon atom (B = 0), no hydrogen atoms (C = 1), one fluorine atom (D = 1), and three remaining bonding positions for chlorine.

$H_2CF\text{-}CF_3$ is called R-134a. It contains no double bonds (A = 0), two carbon atoms (B = 1), two hydrogen atoms (C = 3), and four fluorine atoms (D = 4). The letter "a" indicates that three fluorine atoms are at the same carbon atom.

$HCCl=CH\text{-}CF_3$ is called R-1233zd (E). It contains one double bond (A = 1), three carbon atoms (B = 2), two hydrogen atoms (C = 3), and three fluorine atoms. The free position is occupied by chlorine (one atom). The letters "zd" and "(E)" specify the isomer structure.

Hydrochlorofluorocarbons (HCFCs) were the first alternative blowing agents with reduced ODP, in which at least one halogen atom was substituted by hydrogen. Their ODPs ranged from 0.02 to 0.11. Because improved alternatives became available soon thereafter, they were phased out again after a few decades.

The efforts to further reduce the ODP led to the development of hydrofluorocarbons (HFCs), which only contain fluorine. The HFCs show zero ODP values but significant GWP values ranging from 800 to 2,900. In parallel, PU systems were developed using available non-halogen-containing volatile organic compounds such as hydrocarbons (HCs), methyl formate, and dimethoxymethane. These blowing agents show zero ODP and low GWP values. Because of their low molecular masses, they exhibit relatively high heat conductivity values (λ-value or k-factor).

The most recently developed blowing agent class is that of hydrofluoroolefins (HFO). These novel blowing agents exhibit zero ODP, low GWP values, and excellent thermal conductivity properties. Tab. 2.7 summarizes the ODP, GWP, and heat conductivity values of the most commonly used blowing agents in the past and present. The blowing agents in today's PU applications show zero ODP and low GWP values.

Tab. 2.7: Physical blowing agents and their properties.

Blowing agent	Substance class	Molecular mass (g/mol)	Boiling point (°C)	λ-value at 25 °C (mW/(m·K))	ODP (R-11 = 1)	GWP (CO_2 = 1)
R-11	CFC	137.5	23.8	8.7	1	4,660
R-30 (CH_2Cl_2)	HCC	85	39.6		0	8.7
R-22	HCFC	86.5	−40.8	11.0	0.055	1,760
R-141b	HCFC	117	32	9.7	0.12	725
R-134a	HFC	102	−26.3	13.6	0	1,430
R-245fa	HFC	134	15	12.2	0	1,030
R-227ea	HFC	170	−16.4	12.5	0	3,220
R-365mfc	HFC	148	40.2	10.6	0	794
R-1233zd(E)	HCFO	130.5	19	10.5	0	6
R-1336mzz(Z)	HFO	164	37.4	10.7	0	9
n-Pentane	HC	72	36	14.0	0	4
Isopentane	HC	72	27.7	13.6	0	4
Cyclopentane	HC	70	49	12.4	0	10
Methylal		76	42.3	15	0	<3
Methyl formate	HC	60	32	10.7	0	<5
CO_2		44		16.0	0	1

The volume of gas generated per mole of the blowing agent is nominally a constant. Thus, with the increasing molar mass of the blowing agent, more blowing agent is required to achieve the same foam density. Low-molecular-mass blowing agents are, therefore, by weight, more efficient in expanding the foam.

The boiling point of the blowing agent is critical for processing. When the blowing agent is liquid at room temperature, it can be easily dosed. Gaseous blowing agents must be processed under pressure and exhibit rapid foam expansion when the liquid leaves the nozzle of the mixing head. So-called frothing techniques must be applied when using these blowing agents.

Flexible foam is commonly expanded with carbon dioxide from the isocyanate-water reaction. If further reduction of the foam density is required, gaseous carbon dioxide can be applied, but it requires an adapted mixing machine.

Most rigid foam applications use a combination of chemical and physical blowing to optimize performance, cost, and environmental impact. The choice of the physical blowing agent depends on the type of application. Nowadays, n-pentane, isopentane, and cyclopentane are the most commonly applied blowing agents in the manufacturing of lamination boards and sandwich panels. Refrigerators and freezers in Europe are predominantly expanded with cyclopentane; however, in the United States, most producers use HFCs and HFOs.

2.7.2 Surfactants

Silicone-based surfactants stabilize the foam cells during the foaming process. These surfactants are grafted copolymers with a polydimethylsiloxane backbone and pendant polyalkylene oxide groups.

The most important property of the surfactants is that they lower the surface tension at the liquid-liquid and liquid-air interfaces. During mixing, they enhance the emulsification of the incompatible polyurethane components and facilitate the nucleation of the reacting liquid with air bubbles. The surfactants stabilize the tiny air bubbles formed during mixing and the expanding cells during foam growth.

The surfactants must be at least partially soluble in the polyurethane formulation and have low surface energy structures. The low surface energy part comprises the polydimethylsiloxane chain, whereas tethered polyether chains achieve compatibilization with the PU formulation. The surfactant is synthesized by hydrosilylation of allyl-ended polyether chains with Si-H-containing polysiloxanes in the presence of platinum catalysts.

The following parameters can be adjusted to achieve the correct balance between blend compatibility and surface activity:
- Molar mass of the polydimethylsiloxane and polyether chains
- Ratio of siloxane to polyether side chains (a:b ratio)
- Ratio of EO/PO in the polyether side chain (n:m ratio)
- Reactivity of end groups – R can be hydrogen or alkyl

Numerous surfactants are available for the various PU foam applications to achieve optimum foam stability.

2.7.3 Flame retardants

Fire retardants are added to polymers to reduce their inherent flammability. The chemical or physical reactions induced by the fire retardant slow or stop the spread of the flame and reduce the burning intensity.

Halogen-containing fire retardants act as radical scavengers, interrupting gas-phase radical reactions. Bromine-containing compounds are effective in reducing the flammability of rigid foam. The three bromine-containing fire retardants listed below contain hydroxyl groups, allowing for covalent bonding to the PU polymer. This renders the toxicology of the additives in foams unproblematic.

Phosphorus additives enhance char formation, which protects the material beneath from burning by shielding the heat and forming a gas diffusion barrier. The most commonly used flame retardant in the PU industry is tris(2-chloropropyl) phosphate (TCPP), which contains the elements chlorine and phosphorus. The most commonly applied non-halogen phosphorus-containing fire retardant is triethyl phosphate (TEP).

Tab. 2.8 presents the main phosphorus and bromine fire retardants currently employed, along with the percentages of fire-effective elements in each compound.

Tab. 2.8: Flame retardants.

Flame retardant	Molecular mass (g/mol)	Phosphorus (wt.%)	Chlorine (wt.%)	Bromine (wt.%)	Compound
TEP	182	17.0	–	–	(1)
DEEP	166	18.7	–	–	(2)
TCPP	327.5	9.5	32.5	–	(3)
TPP	326	9.5	–	–	(4)
RDP	574	10.8	–	–	(5)
Tetrabromo phthalate diol	≈ 628	–	–	≈ 51	(6)
Dibromo neopentyl glycol	262	–	–	61	(7)
Brominated polyether triol	≈ 477	–	≈ 15	≈ 33	(8)

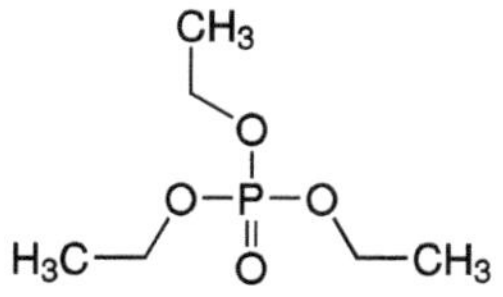

triethylphosphate (TEP)
(1)

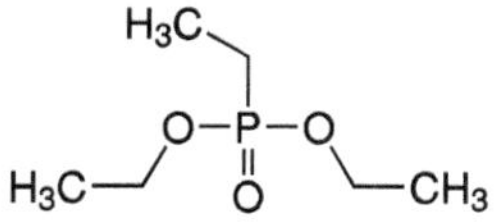

diethyl ethylphosphonate (DEEP)
(2)

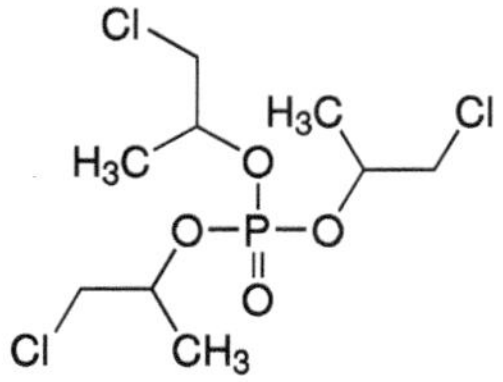

tris(chloroisopropyl)phosphate (TCPP)
(3)

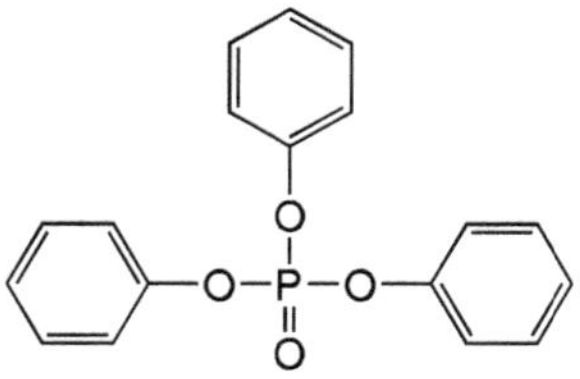

triphenylphosphate (TPP)
(4)

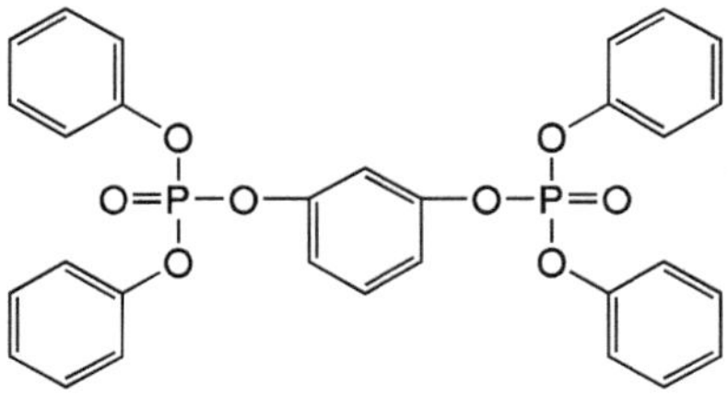

resorcinol bis(diphenylphosphate) (RDP)
(5)

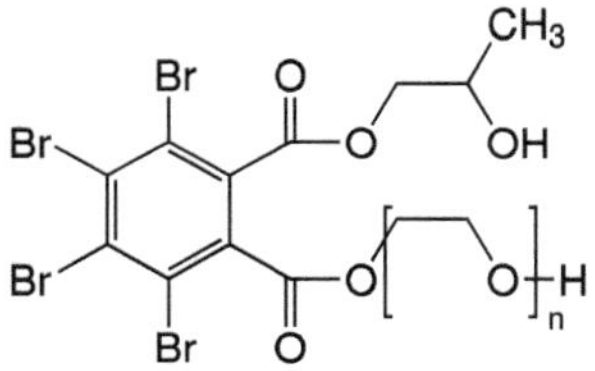

tetrabromo phthalate diol
(6)

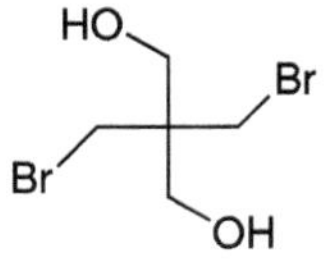

dibromo neopentyl glycol
(7)

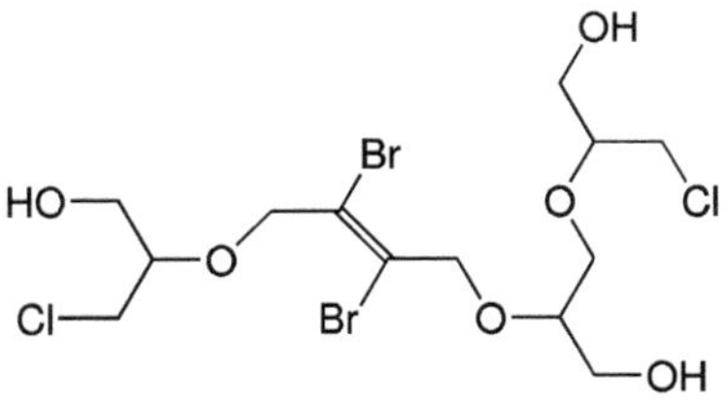

brominated polyether triol
(8)

Solid fire retardants are also used in PU. Melamine and expandable graphite are used in flexible foams, while magnesium hydroxide and aluminum oxide trihydrate are utilized in compact polyurethanes, such as TPU. Ammonium polyphosphate is employed in intumescent coating applications.

References

Further reading

- R.M. Hill, Silicone Surfactants, Marcel Dekker, New York, 1999, ISBN 0-8247-0010-4.
- G. Oertel, Polyurethane, Kunststoff Handbuch (Becker/Braun), Band 7, 1993, ISBN 3-446-16263-1.
- D. Thorpe, D. Sparrow in The Polyurethanes Book, D. Randall and S. Lee, Eds, Wiley, 2002, Chapters 5 and 6, ISBN 0-470-85041-8.
- Ullmann's Encyclopedia of Industrial Chemistry, 7th ed., Wiley-VCH, (2011) ISBN 9783527329434.

Specific references

[1] H. Ulrich, Chemistry and Technology of Isocyanates, John Wiley & Sons, 1996, p. 387, ISBN 0-471-96371-2.
[2] S.F. Stahl, G.A. Luinstra, Analysis of propoxylation with zinc-cobalt double metal cyanide catalysts with different active surfaces and particle sizes, Reaction Chemistry & Engineering, 9, 91–107 (2024)
[3] M. Ionescu, Chemistry and Technology of Polyols for Polyurethanes, 2nd Edition, Volume 1 and 2, United Kingdom, Smithers Rapra Publishing, 2016, ISBN 978-1-910242-14-8 and ISBN 978-1-910242-99-5.

3 Polyurethane chemistry

The term polyurethane is used to designate polymers produced by the polyaddition of polyfunctional isocyanates with molecules that bear at least two hydroxyl groups. The name polyurethane is derived from the urethane group formed in this reaction. More generally, isocyanate reacts with alcohols and other components that contain "active" hydrogen atoms, such as carboxylic acids, amines, and water. Polyurethane materials may, therefore, also exhibit other linkages, such as urea and amide. Furthermore, isocyanates can react with themselves, such as forming isocyanurate.

The high reactivity of the isocyanate is the key to the formation of polyurethane. The PU technology is based on aromatic isocyanates due to their cost-effectiveness, polymer properties, and high reactivity. Polyurethanes are produced using a two-component processing method, where the two components are the polyol blend and the isocyanate. The former contains polyols, additives, and other reactive compounds. The reactions are exothermic, providing sufficient thermal energy to enable rapid curing. Catalysts balance the rate of the various reactions and ascertain high conversions in short reaction times. Physicochemical changes in the reactive mixture may occur during the later stages of the reaction. The reaction can then transition from a chemical to a diffusion-controlled process. The reaction rate slows down, and complete conversion of all the reactive groups may not be achieved.

The structure of the employed polyols and isocyanates determines the properties of the polyurethane. Depending on the polymer morphology and chain topology, polyurethanes can be classified as thermoplastic or thermosetting, as well as elastomeric or rigid. They can be foamed and show a cellular structure or be compact.

3.1 Reactivity of the isocyanate group

Isocyanates have the general structure R–N=C=O and, therefore, contain a carbon atom forming part of an imino and a carbonyl group. The unsaturated carbon atom of the cumulated double bond exhibits a partial positive charge, as evident in the resonance structures of the isocyanate.

$$R-\overset{\ominus}{N}-\overset{\oplus}{C}=O \quad \longleftrightarrow \quad R-N=C=O \quad \longleftrightarrow \quad R-N=\overset{\oplus}{C}-\overset{\ominus}{O}$$

$$(1) \qquad\qquad\qquad \updownarrow \qquad\qquad\qquad (2)$$

$$\overset{\ominus}{R}=N-\overset{\oplus}{C}=O$$

$$(3)$$

The oxygen shows the most substantial net negative charge (2), and the nitrogen has an intermediate net negative charge (1). The positive charge at the carbon atom can

https://doi.org/10.1515/9783110744583-003

be further stabilized when R is an aromatic group. The aromatic ring favors meso-meric structures that stabilize the positive charge at the central carbon atom of the isocyanate. Canonical form (3) gains importance because a resonance-stabilized anion can be generated through the -M effect. This stabilization mechanism is absent in ali-phatic isocyanates. Consequently, aromatic isocyanates are significantly more reactive than aliphatic isocyanates. The reactivity differences between aliphatic and aromatic isocyanates can be as high as two orders of magnitude.

The reaction of isocyanates with components containing active hydrogen groups, denoted as HX, occurs across the C=N bond. The nucleophile X attacks the electro-philic central carbon atom of the isocyanate. Because the adjacent nitrogen and oxy-gen atoms are more electronegative than the carbon atom, the negative charge built up is stabilized. The proton transfer from the active hydrogen component to the nitro-gen atom of the isocyanate completes the reaction.

The reactivity of the isocyanate is primarily influenced by electronic effects resulting from the chemical nature of R, but steric effects also play a role. Electron-withdrawing groups on R will increase the electrophilicity of the carbon atom, thereby increasing the reactivity of the isocyanate toward nucleophilic attack. Conversely, electron-donating groups will destabilize the positive charge and reduce the reactivity. In this context, an isocyanate is an electron-withdrawing group, whereas urethane is an electron-donating group.

The two isocyanate groups in TDI are attached to one aromatic ring; in MDI, each aromatic ring carries one isocyanate.

2,4-TDI   2,6-TDI   4,4'-MDI   2,4'-MDI

The reactivity of the isocyanate in the 4-position of 2,4-TDI is twofold higher than that in 4,4'-MDI, as the second isocyanate activates it. Once the isocyanate in the 4-position has reacted to form urethane, the second isocyanate in TDI in the 2-position reacts about ten times more slowly because the urethane group deactivates it and the methyl substituent in the ortho position sterically hinders it [1]. The intrinsic reactiv-ity of the second isocyanate in MDI in the 4'-position is, in principle, not affected by the presence of the urethane in the 4-position at the other ring. When the first isocyanate has reacted, the second will react at half the rate of the first [2] [Tab. 3.1]. The reactivity of the second isocyanate of MDI is then about twice that of TDI. The reaction rate of the second isocyanate in the diisocyanates is fundamental in building high-molar-mass polymers. Therefore, MDI-based systems are, overall, more reactive than TDI systems and need less catalyst at equal cure rates.

Tab. 3.1: Approximate relative reactivity of the isocyanate groups in 4,4'-MDI, 2,4'-MDI, and 2,4'-TDI.

Position	MDI	TDI
2	0.2	0.2
4	1	2
4'	0.5	

The effect of steric hindrance is important when comparing the reactivity of the two isomers, 4,4'-MDI and 2,4'-MDI. The former is a symmetrical molecule with two groups of equal reactivity. The latter is an asymmetrical molecule with two groups of different reactivities. The reactivity of the isocyanate group in the para-position is approximately four to six times higher than that of the ortho-positioned isocyanate, depending on the reacting alcohol and reaction conditions [3].

3.2 Isocyanate reactions with active hydrogen compounds

Isocyanate can react with compounds containing "active hydrogen" groups, such as hydroxyl, amine, water, and carboxylic acids, forming urethane, urea, and amide groups. Urethane and urea can, in this respect, also be considered as "active hydrogen" compounds yielding allophanate and biuret. In PU systems, the various reactions may run in parallel, the rate of which can be steered by catalysis. Urethane and urea groups possess hydrogen-donating (N-H) and hydrogen-accepting (C=O) sites. The urethane and urea hydrogen bonding interaction is considerable and important for the mechanical properties of PU polymers.

3.2.1 Urethane formation

The most important reaction in polyurethane technology is the urethane reaction, in which an isocyanate reacts with a hydroxyl group to form the urethane linkage. The reaction is exothermic, with an enthalpy of approximately 90 kJ/mol [4]. This reaction links a polyol with an isocyanate, increasing the molar mass of the growing polymer. The urethane reaction is, therefore, also referred to as the gelling reaction.

The nucleophilic oxygen atom of the alcohol attacks the electron-deficient carbon atom of the isocyanate, forming a highly charged intermediate. In a rate-determining step, a proton transfer occurs, resulting in the formation of the urethane bond.

The reaction is reversible; at higher temperatures, the bond dissociates, and the isocyanate and alcohol reform. The temperature at which urethane dissociates depends on the alcohol and isocyanate used. The more acidic the hydrogen atom of the hydroxyl compound and the more reactive the isocyanate, the lower the dissociation temperature of the bond. Phenol is more acidic than alkyl alcohol, and aromatic isocyanates are more reactive than aliphatic isocyanates. The dissociation temperatures of the four possible urethanes are given in Tab. 3.2. The temperatures are approximate; the actual values may differ depending on the measurement method and conditions.

Tab. 3.2: Approximate dissociation temperature (T_d) of the urethane bond [2].

Isocyanate	Alcohol	T_d (°C)
Alkyl	Alkyl	250
Phenyl	Alkyl	200
Alkyl	Phenyl	180
Phenyl	Phenyl	120

Urethanes prepared from aliphatic isocyanates and alkanols are the most stable, followed by urethanes from aromatic isocyanates and alkanols. The urethanes from aromatic isocyanates and phenols are the least stable. The poor stability of the latter is utilized to prepare "blocked" isocyanates for oven-curing coatings. The phenol-blocked isocyanate is used in combination with an aliphatic polyol. The two components are mixed, resulting in a stable coating composition at room temperature. The mixture is then applied and heated to its activation temperature, at which the isocyanate deblocks. The blocking agent evaporates, and the released isocyanate reacts with the aliphatic polyol to build the coating.

3.2.2 Urea formation

The second most important reaction is the reaction of isocyanate and water. The rate-determining step in this reaction is the nucleophilic attack of the oxygen atom of water on the central carbon atom of the isocyanate, resulting in the formation of carbamic acid. Carbamic acid is unstable and readily releases carbon dioxide. The amine that is formed quickly reacts with a second isocyanate to form the disubstituted urea bond. Hence, two moles of isocyanate are required for every mole of water.

carbamic acid

The formed carbon dioxide can be used to expand the foam, which is why this reaction is called the blowing reaction. At the same time, this reaction is a chain extension reaction because it links two isocyanates through a urea linkage. The urea reaction is highly exothermic, with an enthalpy of approximately 135 kJ per mole [4]. The formed urea bond is more stable than the urethane bond from the same isocyanate and an aliphatic alcohol.

The reaction of amine and isocyanate can also be utilized in its own right. Primary and secondary amines react with isocyanate to form di- and tri-substituted urea. Amines are strong nucleophiles, and their reaction to isocyanates is fast. The reaction rate decreases with a decrease in the nucleophilicity of the amine. Hence, aromatic amines are less reactive than aliphatic amines. Steric hindrance and electron-withdrawing groups on the amine further reduce the reactivity. The reaction of an amine with isocyanate exhibits a reaction enthalpy of approximately 100 kJ/mol [4].

3.2.3 Reaction of isocyanate with carboxylic acids

The reaction of a carboxylic acid with an isocyanate first yields an unstable mixed anhydride, which decomposes to form an amide and carbon dioxide. The reaction is relatively slow and is not commonly used in PU applications.

The reaction of isocyanate with formic acid is a special case. It requires two isocyanates for every formic acid molecule, yielding urea, carbon dioxide, and carbon monoxide. The gas yield is high, with one gas molecule per isocyanate. Formic acid is used as a chemical blowing agent in rigid PIR foam.

3.2.4 Hydrogen bond formation

Infrared (IR) spectroscopy is widely applied as an analytical technique to study polyurethane reaction kinetics and structure development. Consider the reaction of an aromatic isocyanate with an alcohol [4]. During polymerization, the isocyanate, exhibiting a strong absorption at 2,270 cm^{-1}, is converted into strongly IR-absorbing groups, such as urethane and urea. The C=O bond in urethane and urea exhibits absorptions at 1,730 and 1,710 cm^{-1}, respectively, and are referred to as "free" urethane and urea. Depending on their molecular structure, environment, and concentration, the urethane and urea groups in PU flexible foams and elastomers can undergo mutual hydrogen bonding. The C=O stretching frequency in the hydrogen-bonded form exhibits substantial shifts due to the additional constraints imposed by hydrogen bonding. The hydrogen-bonded urethane C=O group shifts to 1,710 cm^{-1}. Urea can undergo mono- and bi-dentate hydrogen bonding interactions, with C=O stretching shifts to 1,670 and 1,640 cm^{-1}, respectively [Tab. 3.3].

The urethane and urea N-H also show a shift upon hydrogen bonding. The free N-H in both bonds shows an adsorption peak at 3,450 cm^{-1}, whereas the hydrogen-bonded urethane and urea N-H shift to 3,290 and 3,300 cm^{-1}, respectively. The frequency of the N-H vibrations in mono and bidentate urea hydrogen bonding is the same. The urethane and urea N-H peaks are of low intensity and relatively broad. This may be why N-H vibrations are not systematically used in morphological studies.

3.2 Isocyanate reactions with active hydrogen compounds — **53**

Tab. 3.3: IR shifts of the C=O and N-H bonds in urethane and urea upon hydrogen bond formation.

3.2.5 Allophanate and biuret formation

The N-H in urethane and urea are active hydrogens and can react with isocyanate, giving allophanate and biuret, respectively.

Both reactions are exothermic and reversible at higher temperatures. Uncatalyzed, the allophanate and biuret reactions become significant at temperatures above 100 and 130 °C, respectively. The allophanate reaction reverses at approximately 150 °C, whereas the biuret is more stable. Both reactions can occur during the formation of polyurethane. However, they only become significant when an excess of isocyanate is used. The formation of allophanate and biuret can be desirable because they provide additional chemical crosslinks, which may improve the polymer's physical properties.

3.2.6 Oxazolidinone formation

Oxazolidinone (oxazolidin-2-one) can be formed in the stoichiometric reaction of an epoxide with an isocyanate. The pathway of the three-step reaction leading to oxazolidinone involves the epoxide opening through a nucleophilic attack by an anionic species, followed by the attack of the formed alkoxide on the central carbon atom of the isocyanate. A final ring closure leads to the oxazolidinone through an internal attack of the anionic nitrogen atom on the carbon atom binding the initiating nucleophile. The nucleophile, for example a halide, is liberated upon ring closure.

Efforts were made to find catalyst systems for the selective formation of oxazolidinones. Phosphonium and ammonium halides, such as tetraphenylphosphonium and tetrabutyl-ammonium chloride and bromide, are highly selective catalysts at temperatures of approximately 180 °C. The reaction of bisphenol diglycidyl ether and 2,4-TDI or 4,4′-MDI in the presence of these catalysts yielded linear, high-molar-mass polymers [5]. The thermal stability of oxazolidinones is between that of urethane and isocyanurate.

3.2.7 Reactivity and catalysis

During polyurethane formation, the reactions with the various active hydrogen components and isocyanate will proceed in parallel but with different reaction rates. Amines react faster than alcohols, which, in turn, react faster than carboxylic acids. This shows that the nucleophilicity of the reactive group is more important than its acidity. Furthermore, steric hindrance on the reactive group will reduce its reactivity. Primary alcohols react faster than secondary alcohols, and likewise for amines. The following ranking for amines, alcohols, water, and carboxylic acids is generally accepted (R and Ar stand for aliphatic and aromatic, respectively) [2]:

$$R-NH_2 > R_2-NH > Ar-NH_2 > R-OH \text{ (prim.)}$$
$$> H_2O \geq R-OH \text{ (sec.)} > Ar-OH > RCOOH$$

The reactions of amines with isocyanate are fast and may be finished within seconds. Reactions between isocyanates and hydroxyl groups, as well as with water, are slow and require catalysis. Catalysts accelerate the reaction and regulate the relative rate of the urethane and urea reactions.

Both Lewis acids and Lewis bases can catalyze the nucleophilic addition of active hydrogen compounds to the isocyanate. Lewis base catalysis is broadly applied. Tertiary amines are the most commonly employed Lewis base catalysts. The catalytic activity of amines results from the free electron pair on the nitrogen atom. The tertiary amine can interact with the electron-deficient carbon atom of the isocyanate and the hydrogen atom of the nucleophile. The generally accepted reaction mechanism emanates proton activation – the Farkas and Strohm mechanism [6].

$$R'_3N \ + \ H{-}O{-}R_1 \ \rightleftharpoons \ \overset{\delta+}{R'_3N}{\cdot}{\cdot}\overset{\delta-}{H}{\cdot}{\cdot}O{-}R_1 \ + \ \underset{R_2}{\overset{}{\diagdown}}N{=}C{=}O$$

$$\underset{\underset{\delta+}{R'_3N}{\cdot}{\cdot}H{\cdot}{\cdot}\underset{\delta-}{O}{-}R_1}{\overset{R_2}{\diagdown}N{=}\overset{\delta+}{C}{=}\overset{\delta-}{O}} \quad \xrightarrow{\ -\ R'_3N\ } \quad \underset{H}{\overset{R_2}{\diagdown}}N{-}\overset{\overset{O}{\|}}{C}{-}O{-}R_1$$

The catalyst readily forms an active hydrogen amine complex. The hydrogen bonding interaction increases the nucleophilicity of the active hydrogen component, enhancing the likelihood of addition across the C=N bond of the isocyanate. The catalytic activity of the tertiary amine is governed by its basicity, steric accessibility of the nitrogen atom, and neighboring group participation.

When the tertiary amine structures are similar, the stronger base – the amine with the higher pKa value of the protonated form – will be more reactive. The molecular structures of N,N-dimethylcyclohexylamine (DMCHA) and N,N-dimethylphenylamine (DMPA) are similar; however, their pKa values differ by five orders of magnitude. DMCHA, with a pKa value of 10.7, is a widely used urethane catalyst, notably in rigid foam applications, whereas DMPA, with a pKa of 5.1, shows no catalytic activity. The effect of steric hindrance is illustrated by comparing DMCHA and N,N-diethylcyclohexylamine. Both amines have similar pKa values. The steric accessibility of the nitrogen atom is favorable for the methyl-substituted amine. The methylated amine was found to be a ten-fold stronger urethane catalyst than its more sterically hindered ethylated analog. 1,4-Diazabicyclo[2.2.2]octane (triethylenediamine; TEDA) is a bicyclic tertiary diamine with pKa values of 8.7 and 3.0. For catalysis, only the stronger base is relevant. Despite its lower basicity, it is about four times more catalytically active in the urethane reaction than DMCHA. The bicyclic structure assures low steric hindrance on the nitrogen atom. The amine remains near the active hydrogen atom in the transition state and can efficiently catalyze the rate-determining proton transfer.

Bis(2-dimethylaminoethyl) ether (BDMAEE) is an excellent catalyst for the isocyanate-water reaction. Its structure – a symmetrical molecule containing three hetero atoms spaced by two carbon atoms – allows the formation of stable hydrogen bonds with water through multiple bonding interactions. The strong hydrogen bonding increases the charge density of the water oxygen atom, which facilitates its nucleophilic attack on the isocyanate. This multi-cryptand hydrogen bonding interaction does not occur with hydroxyl groups; instead, urethane catalysis occurs through a straightforward acid-base interaction between the amine and hydroxyl group, as discussed previously. The central oxygen atom, however, withdraws charge from the terminal nitrogen atoms, reducing their basicity – the pKa values of the two amines are 9.6 and 7.9. The reduced basicity of the amines decreases their ability to catalyze the urethane reaction. Consequently, BDMAEE possesses a much stronger catalytic activity for the water-isocyanate reaction than for the urethane reaction.

The differences in urethane-to-urea catalyzing properties of the various catalysts used in the industry can be expressed by their "gel-to-blow" ratio [7]. This factor is 20:80 for BDMAEE, whereas it is 88:12 for TEDA. These two catalysts represent the extremes regarding the blow-to-gel activity of commercially available tertiary amine catalysts. DMCHA, for example, has a gel-to-blow ratio of 73:27.

Organic tin compounds are the most generally employed Lewis acid catalysts. These catalysts show high reactivity and specificity toward the urethane reaction. Dibutyltin dilaurate (DBTDL), for instance, exhibits a gel-to-blow ratio of 97:3. The gel-to-blow ratios of DBTDL, TEDA, DMCHA, and BDMAEE are graphically illustrated in Fig. 3.1.

Catalysts are essential ingredients in PU formulations. Catalytic reactivity and chemical selectivity are important parameters when selecting the catalyst. The former

Fig. 3.1: Gel-to-blow ratios of some standard polyurethane catalysts.

3.2.8 Light stability

Polyurethanes based on aromatic isocyanates are susceptible to yellowing and polymer degradation when exposed to UV or sunlight.

The exact mechanism of photolytic degradation of PU is not known. It is believed to occur in aromatic urethanes via a quinonoid pathway. The urethane bridge oxidizes to a quinone-imide structure. This structure is a strong chromophore, resulting in the yellowing of urethanes [8]. PUs from MDI are more sensitive to yellowing than those based on TDI. The reaction scheme below shows a possible quinone-imide structure that could have been formed from an MDI-urethane under exposure to UV light and oxygen.

Yellowing is primarily an aesthetic issue; the mechanical properties remain largely unaffected upon modest exposure to sunlight. The polymers may degrade and lose properties only after long UV exposure, especially at increased temperatures. The yellowing process, however, can be slowed down by adding UV absorbers and free radical scavengers, but the aging process cannot be avoided.

When selecting the appropriate isocyanate for an application where yellowing may be a potential issue, the following criteria must be considered: the cost and reactivity of the isocyanate, as well as the physical properties and UV aging characteristics of the polymer [Tab. 3.4]. Aliphatic isocyanates are three to five times more expensive than aromatic isocyanates and less reactive. The mechanical properties of aromatic isocyanate-based polyurethanes generally outperform those of their aliphatic counterparts due to the superior stability of their hard domains. The drawback of aromatic urethanes is their tendency to yellow. Technical PU applications generally require fast cycle times, good mechanical properties, and low cost. Aromatic isocyanates are suitable when some yellowing is acceptable, such as in industrial coatings and flooring

58 —— 3 Polyurethane chemistry

applications. Aliphatic isocyanates, such as HDI, are used in applications where light stability is a strict requirement, such as automotive topcoat applications.

Tab. 3.4: Criteria for isocyanate selection.

Isocyanate	Aliphatic	Aromatic
Common products	HDI	MDI
	IPDI	TDI
	$H_{12}MDI$	
Costs	high	low
Reactivity	low	high
Properties	soft	strong
Light stability	excellent	fair

3.3 Isocyanate-isocyanate reactions

Isocyanates can also react with themselves, forming uretidinedione, isocyanurate, or carbodiimide. The first and second are also called dimer and trimer because two and three isocyanates are required to form the bond. The formation of carbodiimide releases carbon dioxide.

3.3.1 Uretidinedione, isocyanurate, and carbodiimide formation

Uretidinedione and isocyanurate are formed through cycloaddition reactions across the C=N bond, resulting in four- and six-membered rings, respectively. Carbodiimide formation occurs via a cyclo-addition reaction across the C=N and C=O bond of two isocyanates. The formed four-membered ring is unstable; it loses carbon dioxide and yields the carbodiimide. The loss of carbon dioxide prevents the back reaction, making this reaction irreversible. These bonds can be formed under polyurethane-forming reaction conditions or by reaction-specific catalysis.

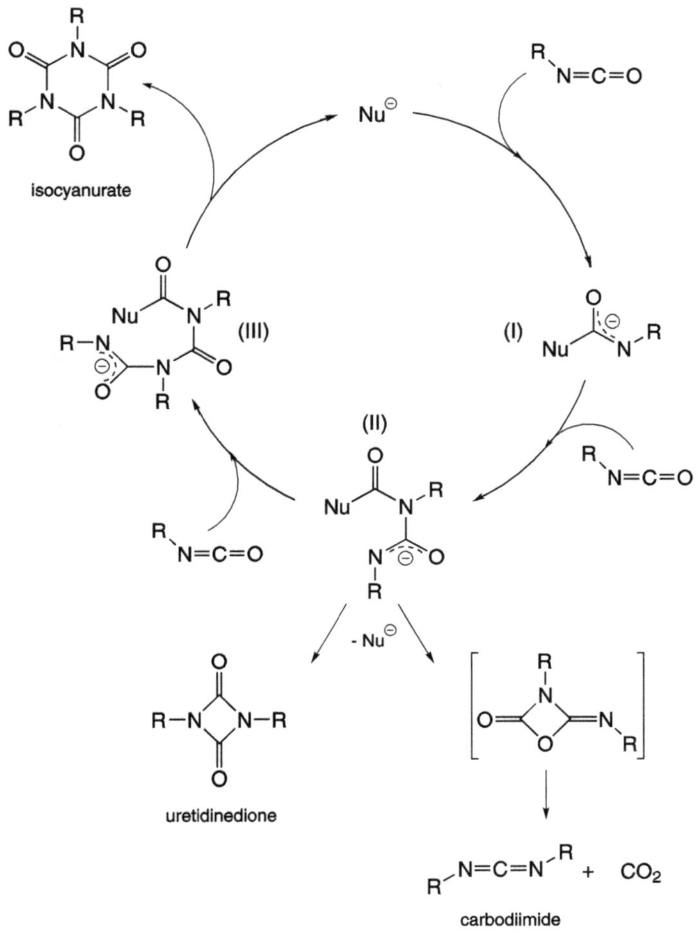

3.3 Isocyanate-isocyanate reactions — 59

isocyanurate

uretidinedione

carbodiimide

The trimerization reaction is important in manufacturing rigid foams (PIR foam) with reduced flammability. When more isocyanate is used than required for the reaction with the polyol, the remaining isocyanate can be converted into isocyanurate. The reaction requires catalysts such as potassium carboxylate and exhibits a high exotherm, which, per NCO group, is approximately half that of urethane formation. Isocyanurate shows high thermal stability and can, in practice, be considered irreversible. Moreover, in a polymer, it forms a crosslink, which further enhances thermal stability. Consequently, polyurethanes containing isocyanurate crosslinks exhibit improved thermal stability and reduced flammability.

The trimerization of aromatic isocyanate in the absence of hydroxyl groups proceeds via an anionic reaction mechanism. A nucleophile attacks the carbon atom of the isocyanate, forming intermediate (I). Similarly, the obtained intermediate (I) can react with two additional isocyanate molecules via intermediate (II) to form (III).

Upon an intramolecular cyclization, an isocyanurate is formed, and the nucleophile is regenerated.

Intermediate (II), however, can also undergo ring closure reactions to yield uretidinedione or carbodiimide, with carbon dioxide as a byproduct. Carbodiimide formation occurs without significant energetic changes, whereas uretidinedione formation is mildly exothermic. The molar distribution of the three reaction products largely depends on the reactant structure, catalyst choice, and reaction conditions. Isocyanurate formation will be dominant when carboxylate catalysts are used. Therefore, carboxylates are used to manufacture polyisocyanurate (PIR) foams, typically using potassium or quaternary ammonium carboxylates.

Uretidinedione formation, or dimerization, can also occur spontaneously in aromatic isocyanate, limiting its shelf life. Dimer formation is especially problematic in isocyanate products that contain high amounts of 4,4'-MDI. The solubility of the dimer in 4,4'-MDI-rich isocyanates is low, in the order of parts per million. When the amount of dimer exceeds the solubility limit, it will precipitate, and the isocyanate becomes turbid. Over time, solids can form, which may impair processing and prevent the use of the isocyanate in production.

Pure 4,4'-MDI can be stored in either liquid or solid form. Either way, the storage conditions have to be chosen so that the dimerization rates are minimized. The dimerization rate is relatively high in the solid phase just below the melting point of 4,4'-MDI, which is 39.5 °C. The molecules in the crystalline phase are aligned with the isocyanate groups positioned in close proximity. Near the melting temperature, the mobility of the molecules in the solid state is sufficiently high to enable dimerization. The rate of dimerization reduces quickly with decreasing temperature. Solid 4,4'-MDI should be stored below 0 °C, preferably below −20 °C, where it can be stored for up to six months. In the liquid phase, crystalline order is no longer present, and concomitantly, the dimerization rate is reduced. A further increase in temperature, however, will increase the dimerization rate, making the best storage condition for liquid 4,4'-MDI just above the melting point, typically between 40 and 45 °C. Isocyanate stored at these temperatures has a storage life of approximately 10 – 15 days, after which it may become cloudy.

If desired, the formation of uretidinedione can be catalyzed using catalysts such as trialkylphosphines and pyridines. The formation of uretidinedione is reversible. It starts dissociating back to isocyanate above 80 °C, but temperatures as high as 190 °C are required to unzip all dimer bonds. Due to the high-temperature requirements, the recovery of a dimerized MDI product by heating is generally impossible.

Carbodiimides, like isocyanates, possess a heterocumulene structure. They can react with active hydrogen compounds, such as water, alcohols, amines, and carboxylic acids [9]. However, the reactivity of carbodiimide with water and alcohols is lower than that of isocyanate. Carbodiimide can form during the polyurethane reaction, particularly at high temperatures and high index. However, the group may not be detected in the final polymer because of subsequent reactions with other components.

Phospholene oxides are effective catalysts for the formation of carbodiimides. The carbodiimide reaction is applied to liquefy 4,4'-MDI. The reaction occurs at temperatures above 100 °C, where the isocyanate is partially converted to carbodiimide. Upon cooling, the carbodiimide undergoes a [2 + 2] cyclo-addition with another isocyanate across the C=N bond, forming a uretonimine. The uretonimine reaction is reversible, with dissociation occurring at approximately 120 °C.

The branched molecule that is formed suppresses the melting point of the isocyanate. Depending on the concentration of uretonimine, a liquid 4,4'-MDI product is obtained at room temperature. Another benefit that uretonimine formation brings to the product is that the solubility limit of the dimer in the isocyanate is increased. Reducing the liquid storage temperature and improving the dimer solubility significantly enhances the shelf life of the uretonimine-modified isocyanate compared to unmodified 4,4'-MDI.

3.3.2 Isocyanurate catalysis

A study on the role of carboxylate catalysts in isocyanurate formation revealed that the actual catalytically active species changed during the reaction, and the carboxylate only served as a pre-catalyst [10]. During the cyclotrimerization of isocyanates, the reaction of the carboxylate anion with isocyanate first forms a mixed anhydride. After the intramolecular rearrangement and decarboxylation of the anhydride, a deprotonated amide is formed irreversibly.

This newly formed anionic species can initiate the polymerization of isocyanate via an anionic polymerization reaction mechanism. Ring closure yields isocyanurate and renders the deprotonated amide.

PIR formulations for the production of rigid foam typically consist of polyol, a three to fourfold excess of isocyanate, and a carboxylate catalyst. The catalysts most frequently applied are potassium acetate and potassium 2-ethylhexanoate. The excess isocyanate is converted into isocyanurate, but this reaction predominantly occurs in the latter stages of the reaction when the temperature is above 60 °C. The isocyanurate formation above 60 °C proceeds fast, which boosts the reaction exotherm. As a result, the foam expansion suddenly accelerates, a phenomenon known as the "second rise" [Chapter 7.3.1].

The catalytic mechanism of isocyanurate formation from isocyanate in the presence of hydroxyl groups, in a ratio typical for producing PIR foam, was studied using potassium carboxylate as a catalyst [11]. Two pathways to isocyanurate have been proposed: the anionic trimerization reaction of isocyanate (Pathway A), as discussed previously, and a consecutive reaction in the presence of alcohol (Pathway B). The latter proceeds via a multi-step reaction pathway whereby the formed urethane is first converted to allophanate, which then reacts to isocyanurate. It was shown that isocyanurate formation proceeded preferentially via the consecutive reaction pathway, whereby allophanate was formed *in situ* in small transient concentrations. Two allophanate-mediated reaction pathways were proposed to form anionic species (III). The allophanate intermediate can undergo an addition-elimination step with (I) yielding (III) and the hydroxyl group. Alternatively, the allophanate reacts with (I) to form (II) and a carbamate. Subsequently, (II) reacts with another isocyanate to form (III). The anionic species (III) cyclizes to form an isocyanurate and liberates the starting nucleophile.

3.4 PU system technology

Polyurethane products are prepared through liquid two-component processing, which involves mixing a polyol and an isocyanate component, referred to as the A- and B-components, respectively. The wide availability of different polyols, isocyanates, and PU additives allows the production of custom-made polymers with a broad spectrum of properties. PU systems are tailored blends that can meet the specific requirements of a particular application. The A-component is a mixture of polyols and additives, such as catalysts, surfactants, pigments, and blowing agents. The B-component represents the isocyanate, which can be a single compound, a mixture of isocyanates, or a prepolymer – the B-component is usually additive-free. Note that the

64 —— 3 Polyurethane chemistry

A and B designations are reversed in the United States. Within this book, we use the European designation.

The chemical industry produces the starting materials, and the PU processors carry out the polymerization to produce PU products. The processor can purchase the starting materials from the open market and utilize its in-house expertise to formulate systems tailored to specific applications. The processor is then referred to as a self-blender. However, a considerable portion of PU starting materials is supplied to the PU producers by the chemical industry in the form of ready-to-use, fully formulated systems. The advantage to the customer of buying fully formulated systems is manifold. Customers can benefit from the chemical industry's extensive knowledge of products and formulations, and the logistics and handling of chemical products can be simplified.

Each system has an optimum isocyanate-to-hydroxy equivalent ratio for the best processing and material properties. This ratio is expressed as the isocyanate index, or, in short, index. The index expresses the equivalent amount of isocyanate used in a formulation relative to the theoretical equivalent amount required.

$$\text{Isocyanate index} = 100 \cdot \frac{\text{amount of isocyanate}}{\text{theoretically required amount of isocyanate}}$$

The isocyanate index is often close to 100, meaning that the number of isocyanate-reactive groups in the A-component and the number of isocyanate groups in the B-component are approximately at stoichiometric ratios. The PU producer, however, can decide to use more or less isocyanate. The implications of this index variation are discussed for a formulation that uses branched polyol and isocyanate to produce a urethane polymer network. When the isocyanate index is less than 100, there is not enough isocyanate to react with all the available hydroxyl groups. The network is incomplete, and network defects in the form of dangling polyol chains will be present. In contrast, at an index above 100, the excess isocyanate groups will react with the urethane linkages, yielding allophanate and forming additional crosslinks.

The PU recipe contains several ingredients, the largest of which are the polyols and the isocyanate. The formulation is generally given in "parts by weight" (p.b.w.) of the ingredients, and can be presented in three different layouts [Tab. 3.5]. In this book, the amount of polyol is taken as 100 parts by weight (p.b.w.), and the weight amounts of other components are then related to this 100 parts of polyol. It provides a quick insight into the used polyurethane ingredients relative to the amount of polyol employed. Another way of presenting the formulation is that the total weight of all compounds in the A-component equals 100 parts by weight. This allows for the facile calculation of the amount of isocyanate, for instance, if index variations are planned. It also allows for calculating the A-to-B ratio, which is important for weighing or metering the required amounts of A- and B-components when preparing, for example, a cup foam or adjusting the machine settings of the dispenser. The third way is to present the amounts of ingredients as a percentage, totaling 100 wt.%. This way, it shows the percentage amounts of all compounds, for example, the fire-retardant percentage, which may indicate the polymer's fire properties.

Tab. 3.5: Comparison of the different layouts to represent a PU system.

Component	Polyol set to 100 p.b.w.	A-component set to 100 p.b.w.	Full recipe (wt.%)
Polyol 1	65	47.8	25.8
Polyol 2	25	18.4	9.9
Polyol 3	10	7.3	4.0
Total polyol	**100**	73.5	39.7
Water	1.5	1.1	0.6
Pentane	10	7.4	4.0
Catalyst	2.5	1.8	1.0
Surfactant	2	1.5	0.8
Flame retardant	20	14.7	7.9
Total A-component	136	**100**	54.0
Isocyanate B	116	85.3	46.0
Sum A and B	252	185.3	**100**

3.4.1 System reactivity

The reactivity of a system is important for processing and productivity. Fast reaction times may favor productivity, but the processing may prohibit high reactivity. The reactivity must be determined before a formulation is tested on the machine. Cup foam testing [Fig. 6.1; page 111] is a daily routine in the lab for reactivity testing and density adjustments of a given system. The reactivity is tested using a small amount of the A- and B-components, typically less than 100 g. The two components are weighed, transferred into a paper cup, and hand-mixed for 5 – 10 s. The following characteristic times are determined, with the start of mixing set as zero.

- Cream time is the period from the start of mixing until the mixture begins to expand.
- String or gel time refers to the point at which the expanding foam begins to gel; experimentally, it is determined by the ability to withdraw polymer strings by dipping and pulling a pin into the growing foam bun.
- Rise time is the time it takes for the foam to reach its maximum height.

The free-rise density can be determined in several ways. The most common method is to cut the foam along the cup's brim and weigh it with its contents. The density is then calculated by abstracting the cup's weight and dividing the foam weight by the cup's volume. The formulator can optimize the reactivity and density using cup foam data before starting machine trials.

3.5 Chain topology and polymer morphology

Depending on the polymer morphology and chain topology, PUs can be classified as thermoplastic or thermoset, as well as elastomeric or rigid. The chemical structure of the polyol and isocyanate predetermines the properties of the resulting polymer. The polyol and isocyanate are characterized by their molar mass and the number of reactive groups on the molecule, known as functionality.

The reaction mechanism of PU formation corresponds to a step-growth polymerization in which a polyol reacts with a polyisocyanate. When both monomers have a functionality of two, linear chains are formed, and the polymer shows thermoplastic behavior. The polymer can be re-melted or dissolved in a good solvent. The melt can be molded into the required shape by standard injection molding. Using monomers with functionality of three or more will yield a crosslinked polymer or thermoset. Thermosets cannot be melted and are produced in their final form. The crosslink density of a thermoset is determined by the functionality of the branches and the molar mass between the crosslinks.

Fundamentally, two different PU polymer morphologies exist. The polymer can be amorphous or show a phase-separated microstructure. Rigid PU, as found in rigid foams and some composite materials, is amorphous and highly crosslinked. Thermoplastic polyurethanes (TPU), elastomers, and flexible foams exhibit a phase-separated structure. These PUs are much softer and possess elastic properties. TPUs have a linear chain build-up, whereas elastomers are loosely crosslinked through covalent bonds. Flexible foams are crosslinked, but their crosslink density is at least ten times lower than that of rigid PU. The chemical flexibility of PU chemistry allows the formation of hybrid structures of these technologies. Still, a given PU product can usually be assigned to one of the following product classes: rigid foam, flexible foam, or elastomers.

3.5.1 Rigid foams

Rigid foams are produced by reacting polyols with molar masses ranging from 200 to 1,000 g/mol, having a functionality between three and eight, with polyisocyanate, typically PMDI. The polymer is highly crosslinked. The crosslink density can be further increased using excess isocyanate, which results in additional crosslinks through allophanate, biuret, or isocyanurate formation. The average molar mass of the chain between crosslinks typically ranges from 200 to 400 g/mol. The polymer is amorphous because the high crosslink density and short branch length of the polyols prevent phase segregation. The polymer can be considered a frozen liquid akin to ordinary glass. The glassy PU polymers are hard and brittle at ambient temperatures but soften at the glass-rubber transition. The glass transition temperature (T_g) of rigid PU increases with increasing crosslink density and generally exceeds 100 °C. Rigid PU can be compact but is usually foamed with densities ranging from 30 to 50 kg/m^3.

3.5.2 Flexible foams

Flexible PU foams exhibit, like elastomers, a phase-separated morphology and elastomeric behavior; however, they must be distinguished from elastomers because they possess different network topologies and hard segment structures. Flexible PU foams are produced by reacting a mixture of a long-chain triol polyether polyols with molar masses between 3,000 and 6,000 g/mol and water, with TDI or MDI. The employed TDI is the standard mixture of 2,4- and 2,6-TDI in a ratio of 80:20 ("T80"), whereas the MDI typically consists of a mixture of 2,4'-, 4,4'-MDI, and polymeric MDI. The urea hard block formed in the reaction of isocyanate and water is incompatible with the polyol soft block, resulting in phase separation. Because mixtures of isocyanates are used, the hard chain segments cannot form ordered hard domains, and consequently, the hard domains are amorphous. Despite the poor stacking, urea-urea hydrogen bonding is observed, which strengthens the hard domains. The hard domains act as physical crosslinks, giving hardness to the foam, whereas the branched monomers provide chemical crosslinks. The average molar mass of the chain between crosslinks in the chemical network ranges from 2,500 to 5,000 g/mol. The carbon dioxide generated in the urea-forming reaction is used to expand the foam. Flexible PU foams are typically produced at densities ranging from 20 to 80 kg/m^3.

3.5.3 Elastomers

PU elastomers are produced through a polyaddition reaction of a mixture of a long-chain diol and a low-molar-mass diol with diisocyanate. The long-chain diol can be a polyether or polyester polyol with a molar mass between 1,000 and 2,000 g/mol. The short-chain diol is also known as a chain extender; the most commonly used is 1,4-butanediol. The isocyanate is preferably aromatic, planar, and symmetric. The reaction yields a linear statistical block-copolymer of the [A-B]$_n$ type. In this formula, A represents the long-chain diol, also known as the soft block. The reaction product of the isocyanate and the chain extender forms the hard chain segment B. The hard chain segments, also referred to as hard blocks, are covalently linked through the long-chain diol. The hard and soft chain segments are incompatible, resulting in a thermodynamically favored phase separation that forms hard and soft domains [Fig. 3.2]. Polymer properties develop through the interaction of hard segments, and the soft phase provides flexibility at low temperatures.

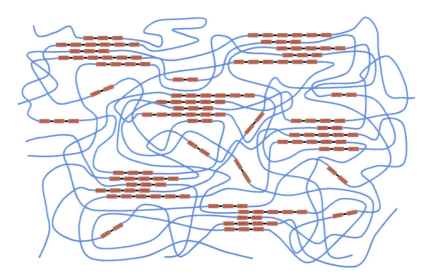

Fig. 3.2: Schematic representation of the phase-separated structure in linear non-crosslinked elastomers.

The regular chemical structure of the hard blocks allows the formation of intermolecular hydrogen bonds. The hard domains of regular structured hard blocks can be partially crystalline. The mechanical properties of the polymer improve with increasing intermolecular interaction of the hard blocks. The hard domains serve a dual role, acting as both physical fillers and physical crosslinkers. They act as fillers, in essence, because the hard domains show a much higher hardness than the surrounding soft phase. At the same time, they act as physical crosslinks – their interconnectivity prevents plastic flow and permanent deformation upon stretching. The soft phase softens at approximately −40 °C, providing the material with low-temperature flexibility. The hard domains melt at approximately 150 °C, at which point the material loses its hardness and mechanical strength. The polymer structure described above is characteristic of TPU. Low levels of chemical crosslinking, as in cast elastomers, do not fundamentally change the polymer morphology but improve high-temperature stability because the covalent crosslinks prevent material flow above the melting temperature of the hard domains. Crosslinked PU elastomers typically exhibit an average molar mass between crosslinks of approximately 5,000 – 10,000 g/mol. They can be compact or foamed, with densities exceeding 200 kg/m^3.

3.5.4 Comparison of rigid and elastomeric behavior

The fundamental property differences between rigid and elastomeric PU are demonstrated by their differences in modulus-temperature behavior at equal density [Fig. 3.3]. Rigid PUs are highly crosslinked polymer glasses and remain rigid till they reach the glass transition. For most rigid PUs, this is approximately 150 °C. Elastomeric PUs exhibit a phase-separated morphology, characterized by hard and soft domains. At temperatures below the glass transition of the soft phase, the moduli of the elastomer and rigid PU are similar; however, they start to deviate at approximately −50 °C, where the soft domains begin to soften, while the rigid PU maintains its modulus. The glass transition of elastomers is broad, spanning a temperature range from approximately −50 to 0 °C. Once the glass transition of the soft phase is surpassed, the modulus remains at a plateau value until approximately 150 °C, at which point the hard domains begin to melt and disintegrate. Despite their fundamental differences, the rigid and elastomeric PUs share the commonality that, within the temperature range of 0 °C to approximately 100 °C, the use temperature of the polymers, the modulus remains approximately constant.

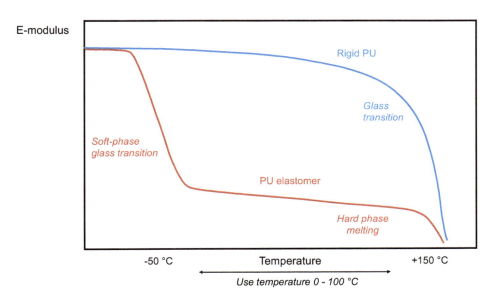

Fig. 3.3: The modulus temperature behavior of elastomeric and rigid PU.

3.5.5 Morphology analysis techniques

Transmission Electron Microscopy (TEM), Atomic Force Microscopy (AFM), and Small-Angle X-ray Scattering (SAXS) can be utilized to investigate the micromorphology of the polymer. Nowadays, AFM is the most commonly applied technique for studying the morphology of PU elastomers. SAXS requires an X-ray source and skills to interpret the scattering results. It is not a common technique but is used in academia to address specific fundamental problems.

Transmission electron microscopy

TEM is a microscopy technique in which a beam of electrons is transmitted through a specimen to form an image. Preparing the samples before measurement requires much effort. The samples must be ultra-thin and require staining to enhance contrast. Staining, however, may impair the results. TEM has lost its attractiveness due to the need for staining and the labor-intensive preparation of samples.

Atomic force microscopy

AFM generates images of a sample's surface by scanning a cantilever equipped with a sharp tip over it. Piezoelectric elements that facilitate accurate tip movements enable precise scanning, and resolutions on the order of fractions of a nanometer can be obtained. AFM can only image a maximum scanning area of about 150 × 150 μm. Therefore, the morphological information obtained is localized. The analysis requires a thin polymer slice with a smooth surface prepared using a microtome.

Small-angle X-ray scattering

SAXS is a scattering technique that allows for the quantification of nanoscale density differences within a material. Because the densities of the hard and soft phases in PU are different, this scattering technique can be used to obtain information on the polymer morphology. The density fluctuations in elastomers cause a beam of incident X-rays to scatter in all directions. From the scattering profile, information can be derived, enabling the determination of the average size and ordering of the hard and soft domains within the polymer. When using powerful X-rays, such as those from a synchrotron, SAXS can provide real-time information on morphological changes in the polymer resulting from tensile deformation, melting, and crystallization.

3.5.6 Elastomer and flexible foam morphology

The phase-separated polymer morphology of a representative PU flexible foam and elastomer was visualized using AFM [Fig. 3.4]. The flexible foam was produced from polyether polyol, water, and a mixture of 2,4'-, 4,4'-MDI, and PMDI, whereas the elastomer was prepared from polyether polyol, 1,4-butanediol, and 4,4'-MDI.

Increasing local hardness changes the arbitrary color scale from dark brown to light brown. The light and dark phases are rich in hard and soft chain segments. The hard domains (light brown) in the flexible foam form a co-continuous structure and are sandwiched between polyol (dark brown) layers. This morphology results from a spinodal decomposition of the hard and soft phases [Chapter 8.3].

The elastomer shows a more complex morphology. It shows small hard domains (white spots, 5 – 10 nm in size) in a continuous matrix consisting of softer (dark brown) and harder (light brown) domains. The small hard domains appear

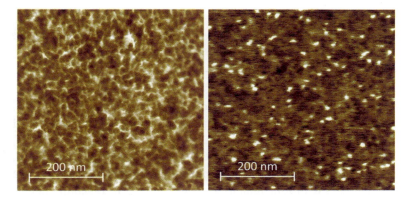

Fig. 3.4: Phase morphology of flexible foam (left) and an elastomer (right) on the same length scale. The AFM picture was taken of a cross-section. Arbitrary color code, white for hard and brown for soft (with the kind permission of BASF SE).

to be relevant for the mechanical strength properties, whereas the polymer morphology of the surrounding material seems more relevant for resilience properties [Chapter 9.6].

In both materials, the phase morphology is characterized by a hard-soft repeating distance (the length of one hard and soft domain) of approximately 10 nm. The soft segment is connected at both ends to two hard segments that are part of two immediate neighbor hard domains. The gap between two hard domains is related to the length of the polyol chain and amounts to approximately 7 nm. The volume fraction of the hard phase in both examples accounts for approximately 30% of the total volume, resulting in a hard domain spacing of approximately 3 nm. The observed repeating distance is then the sum of the two.

3.6 Rheology and cure

The reaction between diol and di-isocyanate gives an alternating copolymer of polyol and isocyanate. The Carothers equation gives the number average degree of polymerization $\bar{P}_n$, where p is the extent of the reaction.

$$\bar{P}_n = \frac{1}{1-p}$$

This relationship indicates that high-molar-mass polymers can only be obtained at nearly complete conversions ($p \rightarrow 1$) [Tab. 3.6].

The relatively slow progression of $\bar{P}_n$ at low values of p can explain the excellent foamability and flowability of PU. At a conversion of 50%, the polymer's molar mass has only doubled. At this point, the reaction exotherm has developed to half its maximum value, while the viscosity increase (related to the polymer molar mass) is still

Tab. 3.6: Degree of polymerization versus conversion rate.

p	0	0.5	0.75	0.9	0.95	0.99	1.0
$\bar{P}_n$	1	2	4	10	20	100	∞

relatively low. The slow viscosity build-up allows easy foam expansion at low-pressure generation. Hence, low-density foams can be produced, and complex mold geometries can be filled.

However, the reaction of a diol and a diisocyanate is not very common in PU. Only TPUs are prepared from two-functional monomers; all other PU products are produced from monomers with functionalities greater than 2. In reaction systems with higher functionality monomers, an infinite degree of polymerization is already obtained at lower conversions. The conversion at the gel point (p_{gel}) can be calculated from the functionality of polyol (f_n-polyol) and isocyanate (f_n-iso), respectively.

$$p_{gel} = \frac{1}{\sqrt{(f_n\text{-}polyol - 1) \cdot (f_n\text{-}iso - 1)}}$$

After passing the gel point, the PU polymer has turned into a loosely crosslinked network. The reaction after gelation occurs both inter- and intramolecularly, resulting in the densification of the network.

The above relates to the rheological changes during PU preparation [Fig. 3.5]. Consider the reaction of a high-functionality polyol and isocyanate, yielding an amorphous rigid polymer. Polyol and isocyanate are low-viscous liquids and can easily be mixed. The reaction begins with mixing the two components, and the viscosity increases. A network forms at the chemical gel point (t_{gel}), the viscosity approaches infinity, and molecular flow over longer distances is prohibited. Further reaction tightens the polymer network, increasing the crosslink density. The mobility of the polymer chains decreases, and the polymer modulus starts to develop. The initial gradual increase of the modulus suddenly accelerates. This is when the polymer starts to vitrify. The time at which the build-up of the polymer modulus accelerates is referred to as solidification time ($t_{solidification}$). Soon after that, the material will have developed sufficient mechanical strength, making it resistant to deformation and allowing it to be handled. This is the handling time ($t_{handling}$). For instance, in a molding application, it means that, at this point, the article can be demolded. The full set of properties will be developed over the next couple of days.

The rheological behavior of elastomer systems looks macroscopically the same, but there are some fundamental molecular differences. PU elastomers are produced in the reaction of a polyol, chain extender, and isocyanate. The functionality of the monomers exhibits a functionality of two or slightly more than two. This implies that chemical gelation occurs at high conversions. Before chemical gelation occurs, however, the polymer physically gels. During the reaction, at some point, the hard segments will reach a critical molecular mass at which phase separation occurs. The hard blocks start to segregate, forming hard domains. The phase separation process, in turn, triggers physical gelation (t_{gel}). More precisely, physical gelation occurs when the glass transition temperature of the hard phases exceeds the reaction temperature. With continued reaction, the hard domains increase in strength, and the modulus increases, showing an S-shaped curve.

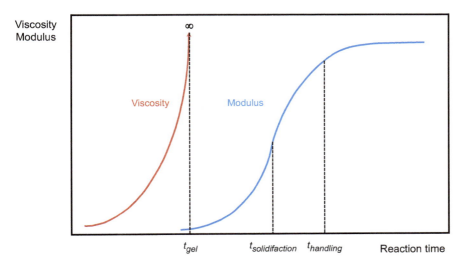

Fig. 3.5: The rheological changes during PU formation. The times at gelation (t_{gel}), solidification ($t_{solidification}$), and handling ($t_{handling}$) are indicated.

3.7 Structure development and reaction rate

The physicochemical transitions that take place during the formation of PU retard the reaction. The slowdown in reactivity can be substantial, even to the extent that full conversion may not be achieved. A complete reaction is desired because only then the full set of mechanical properties can be achieved. The reaction temperature and the time given determine the conversion of the polymerization.

3.7.1 Diffusion control of the reaction

During polymerization, gelation, phase separation, and vitrification may occur [Fig. 3.6]. At each transition, the reaction rate reduces. The reaction rate is the highest at the start of the reaction when the viscosity of the monomer mixture is low and the mobility of the reactive groups is high. The monomers react to larger oligomers, and the intrinsic rate constant (k_1) determines the reaction rate. The reaction at this stage obeys "the principle of equal reactivity", as introduced by Flory: the reactivities of all functional groups are equal to each other, irrespective of the size of the oligomer molecules to which they belong.

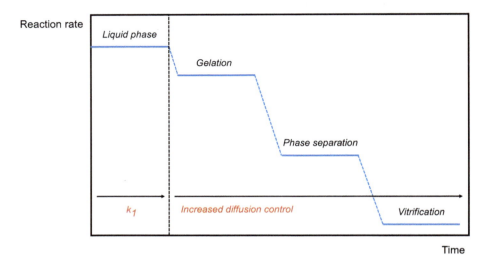

Fig. 3.6: A schematic representation of the changes in reaction rate during the polymerization. A reduction in rate occurs at gelation, the onset of phase separation, and vitrification.

As the reaction progresses and changes in physicochemical conditions occur, the reaction rate slows down because the reaction becomes diffusion-controlled [Fig. 3.6]. The influence of gelation on the reaction rate is comparatively small. The impact of phase separation on the reaction rate depends on the conversion at which phase separation occurs. When this happens at an early stage of the reaction, the impact will be stronger than when it happens at higher conversions. Vitrification has the strongest impact on reactivity. When the material turns into glass, the mobility of the reactive groups gets severely restricted, and the reaction may eventually grind to a halt.

A more quantitative description of a diffusion-controlled reaction is offered [12]. Let us consider a kinetic scheme that includes a diffusion step of the reactants.

$$R-NCO + R_1-OH \underset{k_{-d}}{\overset{k_d}{\rightleftarrows}} [R-NCO \bullet R_1-OH] \overset{k_1}{\longrightarrow} R-\underset{H}{N}-\overset{O}{\underset{\|}{C}}-O-R_1$$

The hydroxyl and isocyanate groups come together via diffusion to form an encounter pair (R-NCO • HO-R) with a rate constant of k_d. This pair can react with a rate constant of k_1 to form urethane. However, the encounter pair can dissociate again with a rate constant k_{-d}. The molecular mobility at the beginning of the reaction is high, and the values of k_d and k_{-d}, both related to diffusion, will be significantly larger than k_1. Consequently, the second step with rate constant k_1 will be rate-determining, and diffusion effects can be neglected.

$$\frac{d[\text{urethane}]}{dt} = k_1 \cdot \frac{k_d}{k_{-d}} \cdot [\text{R}-\text{OH}] \cdot [\text{R}-\text{NCO}]$$

As the reaction proceeds, the molecular mass of the growing oligomer chains will increase, and their mobility will decrease. With that, the absolute values of k_d and k_{-d} will steadily decrease. The reaction may eventually reach a point where k_1 becomes larger than the diffusion rate constants. From then onwards, the reaction becomes diffusion-controlled, and the reaction rate is determined by k_d.

$$\frac{d[\text{urethane}]}{dt} = k_d \cdot [\text{R}-\text{OH}] \cdot [\text{R}-\text{NCO}]$$

Once the reaction is diffusion-controlled, the reaction rate will continue to decrease due to further restriction of the polymer mobility. The impact of diffusion restrictions on the polymerization rate can be visualized in a plot of conversion versus time [Fig. 3.7]. The reaction starts chemically controlled. After gelation, the molecular mobility may be reduced, but the overall reaction rate remains largely chemically controlled. However, the chain mobility decreases when solidification sets in, and the reaction becomes diffusion-controlled.

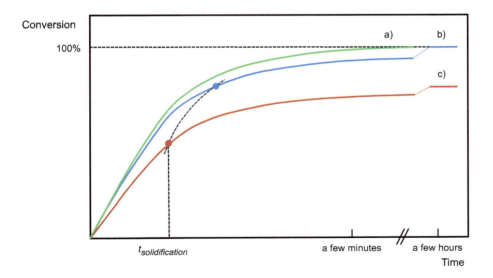

Fig. 3.7: The effect of polymer structure development on the conversion of the reaction for three possible cases: The solidification process (a) does not occur; (b) sets in relatively late; (c) sets in early.

Three cases can be distinguished:
a) The reaction remains chemically controlled until the end of the reaction and achieves complete conversion *(p = 1)* quickly.
b) The reaction is first chemically controlled, and at relatively high conversions, it becomes diffusion-controlled. The reaction may not yet be complete at, for example, demolding time, but it reaches completion with sufficient reaction time.
c) As b), however, diffusion control starts at a relatively low conversion. The reaction slows down progressively and eventually comes to a halt, preventing complete conversion.

A graphical representation of the reaction rate versus conversion for these three cases is shown in Fig. 3.8. The reaction rate develops as follows:
a) The reaction rate is unaffected by the polymer molecular mass and proceeds at a constant rate till full conversion. The polymerization is kinetically controlled and obeys the principle of equal reactivity.
b) Initially, the reaction is chemically controlled; however, the reaction rate decreases as solidification begins. Beyond this point, the principle of equal reactivity no longer holds. The reaction continues, albeit slower, and with sufficient reaction time, complete conversion will be achieved.
c) As b) but now the reaction stops prematurely, and the cure is incomplete.

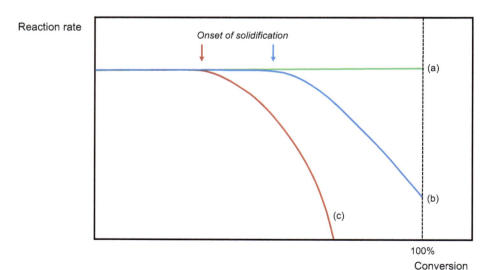

Fig. 3.8: Polymer structure development and its effect on the reaction rate. The solidification process: (a) does not occur; (b) sets in relatively late; (c) sets in early.

3.7.2 Time-temperature-transformation

The curing behavior of glassy PU systems at temperatures above and below the polymer's glass transition temperature can be understood using the Time-Temperature-Transformation (TTT) cure diagram, originally developed by J.K. Gillham [13, 14]. The cure diagram illustrates the relationship between curing temperature and curing time in the context of the physical transformations – gelation and vitrification – that occur during the reaction.

A schematic TTT cure diagram of a rigid thermosetting PU system is shown in Fig. 3.9. The diagram is divided into five zones separated by three contour lines. The contour lines represent gelation, full cure, and vitrification. Each zone represents a material with a different molecular and physical state, including liquid, sol/gel rubber, incompletely cured glass, fully cured rubber, and fully cured glass. The three contour lines can experimentally be determined by measuring the times to gelation, vitrification, and full cure at different isothermal curing temperatures, T_{cure}. Gelation is the point at which the first molecules with infinite molecular mass are formed. Vitrification occurs when the T_g of the reacting polymer exceeds T_{cure}, and full cure $(p = 1)$ corresponds to complete conversion of the reactive groups. $T_{g\infty}$ is the glass transition temperature of the fully cured polymer.

Analogous to the above, the following cases can be considered:

a) $T_{cure} > T_{g\infty}$: When the reaction is carried out at isothermal conditions above $T_{g\infty}$ it will be largely chemically controlled, and full conversion will be achieved quickly. The fully cured rubber obtained at T_a can be turned into a glass by cooling it below $T_{g\infty}$.

b) $T_{cure} < T_{g\infty}$: When T_{cure} is not too far below $T_{g\infty}$, indicated by T_b, the reaction may continue in the vitrified state. The reaction rate in the vitrified state will be reduced; however, a full cure can eventually be achieved with sufficient reaction time.

c) $T_{cure} \ll T_{g\infty}$: When T_{cure} is substantially below $T_{g\infty}$, indicated by T_c, the reaction will strongly decelerate when entering the vitrified state and then freeze (t_{freeze}). Not all the reactive groups have reacted by the time the reaction stops. An incomplete cure of the polymer glass gives network defects, which can lead to brittleness.

Unlike the isothermal reaction conditions in the TTT cure diagram, PU polymerization reactions are generally non-isothermal processes, such as those involved in foaming reactions. With increasing conversion, the heat generated by exothermic processes builds up, the reaction temperature rises, and the foam expands. The core of the rising foam bun becomes self-insulated by the surrounding polymer, which subjects the foaming process to quasi-adiabatic conditions. The foam starts to cool down when the chemical reactions in the foam have ceased, either because vitrification prohibits further reaction or when all the reactive groups have reacted. Despite the non-isothermal reaction

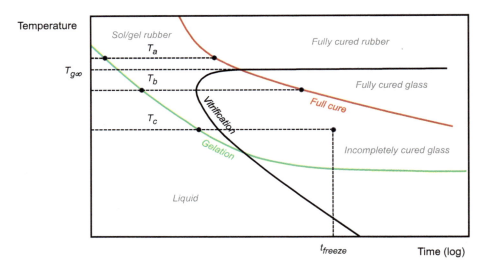

Fig. 3.9: Time-Temperature-Transformation (TTT) cure diagram highlighting the isothermal cure conditions at T_a, T_b, and T_c (adopted from [13]).

process, the TTT cure diagram can be used to qualitatively understand the curing process of rigid PUs [Chapter 7.3.1].

The above theoretical model for forming glassy polymers can, in principle, be extended to phase-separated PUs. However, it is challenging to construct a TTT cure diagram to visualize the relation between phase separation and cure. Fundamental to this problem is that the phase separation is reaction-induced and determined by competing thermodynamic and kinetic (molecular diffusion) factors. For elastomers, a different model is presented, which discusses thermodynamic factors and molecular transport processes seperately [Chapter 9.4.1].

PU elastomers are produced in the reaction of a polyol, chain extender, and isocyanate. Fig. 3.10 illustrates the increase in conversion over time. The conversion at which physical gelation occurs is indicated with a double arrow. Up to this point, the reaction is largely chemically controlled; however, from then onwards, the reaction becomes diffusion-controlled, and the reaction rate slows down. The conversion at which phase separation occurs is critical for the final mechanical properties of the elastomer. In particular, early phase separation should be avoided, as this can lead to a premature termination of the reaction. This may result in poor mechanical properties because high molar mass is not achieved. A remedial measure for this problem can be an increase in the reaction temperature, for instance, by increasing the mold temperature.

With increasing temperature, the hard block aggregation will be subdued. Consequently, the polymer will remain in the liquid state for a longer period, and the conversion at which the reaction becomes diffusion-controlled is delayed. The reaction

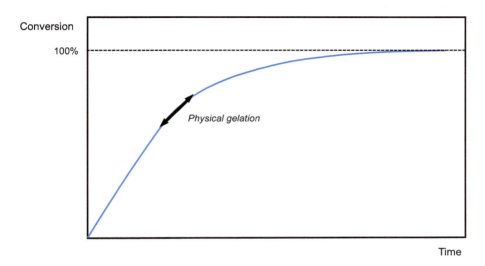

Fig. 3.10: Schematic representation of conversion versus time, highlighting the physical gelation.

can reach higher conversions, and improved mechanical properties can be expected. Increasing the reaction temperature is one option to delay phase separation in elastomers; additional options will be discussed in Chapter 9.4.

References

Further reading

– H. Ulrich, Chemistry and technology of isocyanates, Wiley, 1996, ISBN 978-0-471-96371-4.

Specific references

[1] J. Bosman, R.I. Zimmerman, The Polyurethanes Book, D. Randall and S. Lee, Eds, Wiley, 2002, ISBN 0-470-85041-8, Chapter 7 and 9.
[2] M.F. Sonnenschein, Polyurethanes: Science, Technology, Markets, and Trends, John Wiley & Sons Inc., 2015, ISBN 978-1-118-73783-5, Chapter 3.
[3] L. Nagy, T. Nagy, Á. Kuki, M. Purgel, M. Zsuga, S. Kéki, Kinetics of Uncatalyzed Reactions of 2,4'- and 4,4'-Diphenylmethane-Diisocyanate with Primary and Secondary Alcohols, Int. J. Chem. Kinet., 49 (9), 643–655 (2017).
[4] L.D. Artavia, C.W. Macosko in Low density cellular plastics, N.C. Hilyard, A. Cunningham, Eds, Chapman and Hall, London (1994), ISBN 0 412 58410 7, Chapter 2.
[5] A. Prokofyeva, H. Laurenzen, D.J. Dijkstra, E. Frick, A.M. Schmidt, C. Guertler, C. Koopmans, A. Wolf, Poly-2-oxazolidones with tailored physical properties synthesized by catalyzed polyaddition of 2,4-toluene diisocyanate and different bisphenol-based diepoxides, Polym. Int., 2017, 66, 399–404.

[6] A Farkas, P.F. Strohm, Mechanism of the amine catalyzed reaction of isocyanates with hydroxyl compounds, Ind. Eng. Chem. Fundam., 4(1), 32, 1965.
[7] R. van Maris, Y. Tamano, H. Yoshimura, K.M. Gay, Polyurethane Catalysis by Tertiary Amines, J. Cell. Plast., 2005, 41, 305.
[8] D. Rosu, L. Rosu, C.N. Cascaval, IR-change and yellowing of polyurethane as a result of UV irradiation, Polym. Degr. Stab., 94 (4), 2009, 591–596.
[9] H. Ulrich, Chemistry and technology of carbodiimides, Wiley, 2007, ISBN 9780470065105.
[10] Y. Guo, M. Muuronen, P. Deglmann, F. Lucas, R.P. Sijbesma, Ž. Tomović, Role of Acetate Anions in the Catalytic Formation of Isocyanurates from Aromatic Isocyanates, J. Org. Chem., 2021, 86, 5651–5659.
[11] A.A. Nabulsi, D. Cozzula, T. Hagen, W. Leitner, T.E. Müller, Isocyanurate formation during rigid polyurethane foam assembly: a mechanistic study based on in situ IR and NMR spectroscopy, Polym. Chem., 2018, 9, 4891–4899.
[12] G. Challa, Polymer Chemistry, Ellis Horwood, 1993, ISBN 0-13-489691-2, Chapter 3.
[13] J.B. Emms, J.K. Gilham, Time-Temperature-Transformation (TTT) Cure Diagram: Modeling the Cure Behavior of Thermoset, J. Appl. Polym. Sci., 28 (8), 2567–2591, (1983).
[14] M.T. Aronhime, J.K. Gillham, Time-Temperature-Transformation (TTT) Cure Diagram of Thermosetting Polymeric Systems, in Epoxy Resins and Composites III, Adv. Polym. Sci., 78, 83–113 (1986).

4 Physical properties and flammability

Rigid polyurethane is highly crosslinked and amorphous, exhibiting a glass transition temperature above 100 °C. The glass transition temperature (T_g) must be high so the material is rigid at the use temperature. PU elastomers, including flexible foams, show a phase-separated structure with amorphous soft and glassy or semi-crystalline hard domains. The T_g of the soft phase is approximately -30 °C, whereas the hard domains melt (T_m) at around 150 °C. The T_g must be low to ensure good low-temperature elasticity, whereas the melting temperature T_m determines its high-temperature stability. These transitions and the temperatures at which they occur determine most of the physical properties of the polymer.

PU elastomers are viscoelastic materials. Viscoelasticity is the material property that exhibits viscous and elastic characteristics when undergoing deformation. The presence of the two respective elements is reflected in the elastic behavior of the elastomer – its elasticity, or resilience, increases when the fraction of the viscous elements decreases. The viscoelastic behavior can be described by simple models consisting of viscous dampers and elastic springs. Various experimental techniques are available to determine the T_g and T_m and the viscoelastic properties of elastomers.

PU products may require flammability specifications, such as rigid foams used in the construction industry and flexible foams used in domestic upholstery. The flammability of PUs can be improved by selecting the appropriate starting materials and reaction chemistry, as well as by adding fire retardants. Fire testing is application-specific, but some general basic reaction-to-fire tests exist.

4.1 General thermal behavior of polymers

Small molecules can exist in solid, liquid, or gaseous states. The phase transitions between these three states are sharp and associated with a thermodynamic equilibrium. The phase changes are first-order transitions and involve latent heat. During such a transition, a system either absorbs or releases a fixed amount of energy per unit volume, and enthalpy, entropy, and volume exhibit a sudden change.

The phase behavior of polymer molecules is much more complex. Polymers can be amorphous or partially crystalline, meaning that crystalline and amorphous polymer phases co-exist. The typical states of polymers are glass, rubber, and semi-crystalline, which are thermodynamically metastable. Amorphous polymers exhibit a glass transition, while partially crystalline polymers display both a glass transition and a melting transition. The viscosity of the polymer melt is very high, and chain mobility is low, limiting conformational changes in the chain at the transition temperatures. The glass and melting transitions cannot be strictly considered thermodynamic

https://doi.org/10.1515/9783110744583-004

transitions because of the inherent difficulty in reaching equilibrium near the transition temperatures.

The melting of the crystalline domains in a polymer resembles a thermodynamic first-order transition but differs from that of low-molecular-mass crystals. Polymers do not exhibit a sharp melting point; instead, melting occurs over a temperature range. Furthermore, the melting depends on the specimen history and the heating rate.

The glass transition occurs within amorphous polymers or the amorphous phase of semi-crystalline polymers, where the material transitions from a glassy state to a rubbery state through heating. The glass transition is a continuous phase transition that exhibits features characteristic of a thermodynamic second-order transition. The increased mobility at the glass-rubber transition leads to an increase in the coefficient of thermal expansion, heat capacity, and compressibility. The polymer modulus, which is inversely related to compressibility, decreases above the glass transition temperature. The increase in thermal expansion and heat capacity, along with the decrease in modulus, enables the characterization of the glass transition. The glass transition is kinetically controlled and depends on the rate of temperature change; the higher the heating rate, the higher the T_g. Furthermore, the changes are less abrupt. Fig. 4.1 provides a schematic overview of the changes in volume, polymer modulus, and heat capacity at the T_g and T_m for a fully amorphous and crystalline polymer during a temperature scan. For all practical purposes, however, it can be said that every polymer is characterized by its own T_m and T_g, the values of which are determined by the polymer's chemical structure.

The hard blocks in PU elastomers are the reaction products of a diisocyanate and a chain extender, typically an aliphatic diol such as 1,4-butanediol. The T_m of the hard domains is determined by the stiffness and polarity of the hard blocks and their ability to stack. Stiff chain segments, for instance, aromatic groups introduced via the diisocyanate, increase the T_m. The urethane and urea groups in the main chain can form intermolecular hydrogen bonds, stabilizing the hard domains and thereby increasing the T_m. The strength of hydrogen bonding and other inter-chain secondary interactions depends on the spatial geometry of the hard blocks, which allows them to stack and align. The packing of the hard domains is determined by the molecular structure of the diisocyanate and the chain extender. The planarity and symmetry of the isocyanate and the number of carbon atoms in the aliphatic diol determine the ability of the hard domains to stack, and accordingly, T_m [Chapter 9.3.4].

The most critical factor determining the T_g is the chain flexibility, which depends on the nature of the chemical bonds that comprise the polymer main chain. Incorporating linkages, such as $-O-$ or $-CH_2-$ that allow for easy rotation about the bond axis reduces stiffness, whereas inserting stiff units, such as aromatic groups and urethane linkages, increases chain stiffness. Side groups on the main chain increase the stiffness because they restrict the rotation of the bond. Large and bulky side groups give a strong chain stiffening, whereas the effect may not be so pronounced for small and

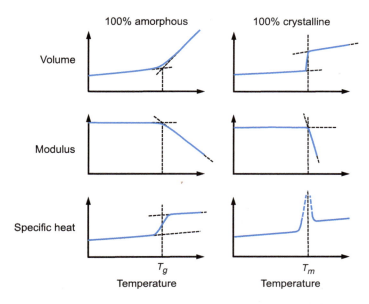

Fig. 4.1: The changes in volume, polymer modulus, and specific heat at transition temperatures T_g and T_m for a fully amorphous and crystalline polymer (by analogy with [1]).

flexible side groups. Polar side groups increase chain stiffness more than non-polar groups of similar size because the additional polar interactions further restrict chain rotation.

Tab. 4.1: Glass transition temperatures (T_g) of polyols [1].

Polyol	T_g (°C)
Poly(ethylene oxide)	-47
Poly(propylene oxide)	-73
Poly(trimethylene oxide)	-62
Poly(tetraethylene oxide)	-83
Poly(ethylene adipate)	-55
Poly(ethylene dodecate)	-71
Poly(decamethylene adipate)	-60

The polyols used in the production of PU elastomers primarily consist of aliphatic polyethers and polyesters. These polyols are highly flexible and exhibit low T_g values [Tab. 4.1]. The most commonly applied polyether polyols, poly(propylene oxide) and poly(tetramethylene oxide), exhibit T_g values of -73 and -83 °C, respectively. The polyester polyols for elastomer applications are adipate-based and exhibit higher T_g values, typically ranging from -55 to -60 °C. The individual polyols have a low and narrow

T_g; however, the T_g of the corresponding soft phase in a PU elastomer is broad. The transition typically begins at -50 °C and reaches 0 °C. The increase and broadening of the T_g result from reduced soft segment mobility exerted by the attached hard chain segments at both ends of the polyol that are part of neighboring hard domains.

4.2 Viscoelasticity

Polymers are viscoelastic materials. The behavior of viscoelastic materials differs from that of pure elastic materials, such as metals. Metals obey Hooke's law; the stress is proportional to the strain, and there is no heat dissipation when a load is applied and removed. The stress-strain properties of viscoelastic materials depend on the strain rate, and heat is dissipated during deformation. During a loading cycle, hysteresis is observed, and the area of the hysteresis loop is directly proportional to the energy lost. A comprehensive knowledge of the viscoelastic behavior of polymers is required to understand the elastic behavior of PU elastomers under service conditions.

Polymers show a combination of viscous and elastic properties. It is assumed that the deformation of a polymer can be divided into an ideal elastic component, represented by the Hooke element, and a viscous component, represented by the Newton element. The elastic element is represented by a spring that obeys Hooke's law, where ε is the applied strain, σ the stress, and E the Young's modulus [2].

$$\sigma = E \cdot \varepsilon$$

The stress is instantaneous, time-independent, and proportional to the strain. The viscous element is represented by a dashpot, composed of a piston and a cylinder filled with a Newtonian fluid of viscosity η. Its deformation is time-dependent and irrecoverable, dissipating the deformation energy as heat. The viscous behavior can be described as a function of the viscosity and the strain rate $(d\varepsilon/dt)$.

$$\sigma = \eta \cdot \frac{d\varepsilon}{dt}$$

The various models to describe viscoelastic behavior involve different combinations of these basic elements. The Maxwell model consists of a spring and a dashpot in series. It describes the relaxation behavior of polymers when subjected to simple harmonic motion [Fig. 4.2].

Dynamic Mechanical Analysis (DMA) is an experimental technique used to study the modulus and viscoelastic properties of polymers. A sinusoidal strain is applied onto the sample, either in torsion or stretching, and the stress in the material is measured. The sample's temperature and the measurement frequency can be varied. The measurements are generally performed at a frequency of 1 Hz at small strains. A temperature scan provides the polymer modulus versus temperature, from which the glass and melting transitions, as well as the polymer's viscoelastic behavior, can be derived.

4.2 Viscoelasticity

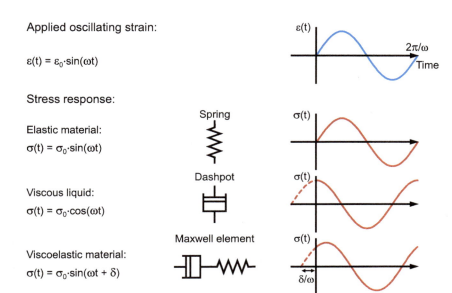

Fig. 4.2: Sinusoidal strain and stress versus time for an elastic, viscous, and viscoelastic material.

The periodic strain in DMA is sinusoidal. Hence the tensile strain varies according to [1, 2]:

$$\varepsilon = \varepsilon_0 \cdot \sin(\omega t)$$

where ε_0 is the amplitude of the sinusoidal tensile deformation, and ω is the angular frequency (2π times the frequency in Hz). The tensile stress for a purely elastic material would vary in phase with the applied strain as:

$$\sigma = \sigma_0 \cdot \sin(\omega t)$$

where σ_0 is the amplitude of the sinusoidal tensile strain. However, for a purely viscous material, the stress would lag $\pi/2$ behind the applied strain as

$$\sigma = \sigma_0 \cdot \sin\left(\omega t + \frac{\pi}{2}\right) = \sigma_0 \cdot \cos(\omega t)$$

A viscoelastic material is neither purely elastic nor viscous, and an intermediate situation arises. The stress lags behind the strain by an angle δ – the "phase angle" or "phase lag". Hence

$$\sigma = \sigma_0 \cdot \sin(\omega t + \delta) = \sigma_0 \cdot [\cos\delta \cdot \sin(\omega t) + \sin\delta \cdot \cos(\omega t)]$$

The stress can be considered as being resolved in two components; the first part is in phase, whereas the second is $\pi/2$ out of phase with the strain. Both parts represent different moduli:

$$E' = \frac{\sigma_0}{\varepsilon_0} \cdot \cos\delta \quad \text{and} \quad E'' = \frac{\sigma_0}{\varepsilon_0} \cdot \sin\delta$$

E' and E'' are the (tensile) storage and loss modulus, respectively. The equation for the tensile stress σ can now be rewritten as:

$$\sigma = \varepsilon_0 \cdot \left[E' \cdot \sin(\omega t) + E'' \cdot \cos(\omega t) \right]$$

E' represents the elastic part of the stress and relates to the recoverable elastic energy stored in the material, whereas E'' represents the viscous part of the stress and relates to the deformation energy dissipated as heat.

The phase angle δ is then given by the ratio of E'' to E':

$$\tan\delta = \frac{E''}{E'}$$

The *tan δ* is the ratio of dissipated energy to the maximum stored energy during a cycle. The quantity *tan δ* is called the "loss tangent" or "loss factor". It is proportional to the internal molecular friction or the specific damping capacity of the polymer, and it is an inverse measure of the rebound resilience. Rebound resilience is "the proportion of the applied energy usefully returned after an impact" (BS 903, Part A8). The resilience of flexible foams is determined by the ball rebound test (ISO 8307). A steel ball is dropped onto a foam test piece from a height of 50 cm, and the rebound height is measured. The rebound resilience is then calculated as a percentage of the rebound height relative to the drop height. Similarly, the resilience of elastomers can be measured using a Schob pendulum (ISO 4662). A pendulum hammer is lifted horizontally and released to fall, striking the vertically mounted test specimen. The rebound resilience of the elastomer can then be calculated from the rebound height of the hammer. The following relationship between resilience and loss factor has been proposed.

$$R = e^{-n \cdot \tan\delta}$$

The relationship is an illustration and can be used qualitatively to relate the rebound and loss factor. Theoretically, n equals π [1].

Like E' and E'', the shear storage and loss modulus, G' and G'' can be defined. The loss angle obtained in both tensile and shear applied tests is similar. Most DMA machines in operation in the PU industry measure in shear, and the DMA examples shown in this book are always measured in shear.

4.3 Dynamic mechanical analysis

Dynamic mechanical analysis (DMA) is used to investigate the effect of polymer composition on the glass and melting transitions, as well as the viscoelastic properties of the PU elastomer. Every transition is characterized by a decline in polymer modulus and is often accompanied by a peak in the loss factor.

Fig. 4.3 shows the DMA results of a TPU produced from poly(tetramethylene ether) glycol, 1,4-butanediol, and 4,4'-MDI. The hard block content (HBC) [Chapter 9.1.4], defined as the mass percentage of chain extender and isocyanate divided by the combined mass of chain extender, isocyanate, and polyol, amounted to 29 wt.%. The polymer sample, produced by injection molding, was annealed at 100 °C for 20 h before further measurements. The G', G'', and $\tan \delta$ are plotted versus temperature, showing the glass and melting transition.

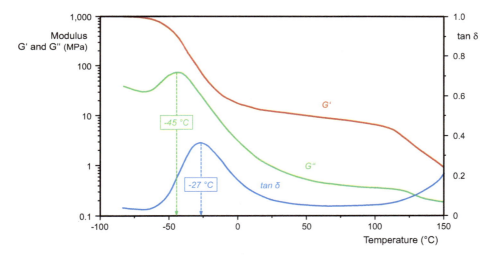

Fig. 4.3: DMA of a polyether TPU with a hard block content of 29 wt.% (with the kind permission of BASF SE).

Below the glass transition, the polymer behaves glass-like; it is hard and brittle. The storage modulus G' of 1,000 MPa is typical for compact amorphous glassy polymers. The polymer begins to soften at approximately -60 °C and then transitions from brittle to ductile. The G' declines whereas G'' increases and reaches a maximum at -45 °C. The high value of G'' at the glass transition indicates that the polymer at these temperatures dissipates high amounts of energy upon deformation; it is tough and robust, like leather. The glass transition is broad and reaches 0 °C. The storage modulus declines over two orders of magnitude during the glass transition. The decrease in G' in this temperature range is primarily attributed to the coordinated movement of the flexible chain segments of the polyol part, which constitutes the more significant mass fraction in the polymer.

A rubbery plateau appears at 0 °C and extends to about 120 °C. The hard phases act as physical crosslinks, and the elasticity results from the stretching of the soft blocks. At 120 °C, the hard domains start melting and lose mechanical stability. The physical polymer entanglement network then takes over the mechanical strength of the polymer, ensuring that the further decline of G' with temperature is relatively smooth.

The temperature at which G'' shows a maximum is generally taken as the T_g of the elastomer. The temperatures at which G'' and the loss factor show a maximum at the glass transition do not coincide. The maximum in G'' marks the point at which the energy loss of the polymer reaches its maximum, whereas the maximum in the loss factor depends on the ratio of G'' to G'. The loss factor maximum always occurs at somewhat higher temperatures than that of G'', and the difference increases with increasing width of the glass transition. The loss factor versus temperature curve is used to study and interpret the damping behavior of the polymer; the G'' curve is usually not interpreted any further.

The loss factor of the present TPU exhibits the lowest values at temperatures between 50 and 100 °C, indicating that the polymer displays the highest resilience at these temperatures. The loss factor at ambient conditions is higher because, at those temperatures, the polymer still operates in the tail of the glass transition. The loss factor increases at temperatures above 100 °C when the hard domains begin to soften. A maximum may not always be observed. This can happen when the sample becomes too soft for proper analysis during the transition.

Fig. 4.4 compares the DMA results of the present TPU with those of a polyether-PMDI-based rigid foam, a polyester-NDI-based foamed elastomer, and a polyether-MDI-based high-resilience flexible foam. All four polymers share a common plateau value of G' from 0 to 100 °C, which corresponds to the working temperature range of the PUs. Below their glass transition temperatures, all polymers are rigid and exhibit similar hardnesses. However, the G' values differ due to their varying densities: the TPU is compact, while the foamed elastomer has a density of 500 kg/m^3, and the rigid and flexible foams have comparable densities of approximately 40 kg/m^3. The rigid foam is a highly crosslinked polymer glass that starts softening at about 110 °C. TPU and flexible foam show relatively broad glass transitions, tailing into the room temperature region. When the hard domains begin to soften, the modulus of the TPU decreases relatively quickly with increasing temperature. The flexible foam, however, largely maintains its modulus because it contains chemical crosslinks that hold the strain. The DMA trace of the NDI-based elastomer is unique; it shows a low T_g and a flat plateau region from 0 to 150 °C. These results can be attributed to the sharp phase separation of the hard and soft domains, as well as the high-temperature stability of the NDI-based hard domains.

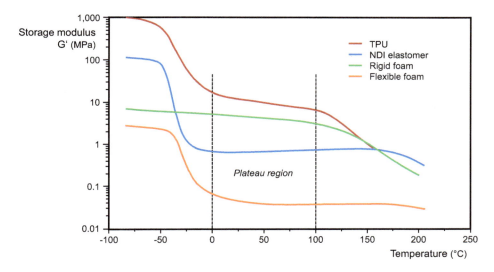

Fig. 4.4: DMA of a TPU, rigid PU foam, NDI elastomer, and flexible foam (with the kind permission of BASF SE).

4.4 Melting of the hard domains

The temperature at which the hard domains begin to soften and the polymer loses its strength can be determined using DMA. The softening is often associated with the melting of the hard domains. The determination of the melting behavior of polymers, however, is complex. Melting occurs over a temperature range and depends upon the conditioning of the polymer, the measurement method, and the rate at which the measurement is performed. The melting enthalpy can be determined using Differential Scanning Calorimetry (DSC). Melting and softening are discussed in the context of morphological changes occurring during these transitions.

4.4.1 Differential scanning calorimetry

Differential scanning calorimetry (DSC) is a thermo-analytical technique in which the difference in the amount of heat required to increase the temperature of a sample and a reference is measured as a function of temperature. The temperature increases linearly with time, and the sample and reference are maintained at nearly the same temperature. The DSC measurement gives a curve of the heat flow (dQ/dt) versus temperature. The glass transition is observed as a sudden increase in heat capacity (c_p), whereas melting the hard phases yields an endothermic effect [Fig. 4.1; page 83].

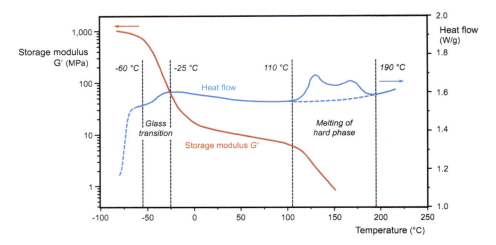

Fig. 4.5: DSC and G' traces from -70 to 220 °C of a polyether TPU with a hard block content of 29 wt.% (with the kind permission of BASF SE).

Fig. 4.5 shows the DSC thermogram of the same polyether TPU used in the DMA experiment above [Fig. 4.3; page 87]. The T_g of the soft phase is observed as an S-shaped jump in c_p starting from -60 °C and reaching -25 °C. The melting transition is broad, spanning 80 °C from 110 to 190 °C, and exhibits maxima at 130 and 170 °C. The first peak is the result of the annealing process and is therefore termed the "annealing peak".

The broad melting range is indicative of a broad distribution of hard domain sizes. The melting temperature of the hard domains increases with their size; small domains melt first, whereas the larger ones survive longer. During cooling from the melt, large domains form first, so the size and melting temperature of the domains are in the reverse order of their formation. TPU solidified from the melt during injection molding contains a relatively small amount of hard phase, about half of what is theoretically possible. The annealing process significantly increases the hard-phase volume because dissolved and poorly arranged hard blocks can reallocate and form new or join existing hard domains. Annealing generates considerable order within and among the hard domains. The phase separation between the hard and soft domains improves, and the T_g decreases. Furthermore, the urethane-urethane hydrogen bonding interaction increases, resulting in improved cohesive strength properties of the hard domain. The hard blocks that were dissolved in the soft phase before annealing were likely the smaller ones, forming the less stable and lower melting hard domains. It is, therefore, conceivable that the smaller hard blocks generate the annealing peak.

DSC detects differences in heat flow during the scan but cannot discriminate between the melting and dissolution of the hard domains. Scattering techniques can provide deeper insights into the polymer morphology and the morphological changes that occur during melting.

4.4.2 X-ray scattering

Ordered hard domains may show crystallinity. When crystalline structures are present in the polymer, they can be detected using Wide-Angle X-ray Scattering (WAXS). The phase morphology can be studied using Small-Angle X-ray Scattering (SAXS).

When X-rays are directed at a partially crystalline material, they are scattered in characteristic patterns from which the crystal structure and the crystalline volume fraction can be calculated. WAXS on PU elastomers can detect all crystallinity relevant to the material properties. The nano-scale morphology of hard and soft domains in PU elastomers can be studied using SAXS. The SAXS technique is similar to WAXS, except that the distance from the sample to the detector is longer, and diffraction maxima are detected at smaller angles. The density differences between the hard and soft phases cause the X-rays to scatter, which SAXS detects. Increasing density contrast increases the scattering intensity and *vice versa*. SAXS and WAXS spectra can be recorded simultaneously to obtain information on the crystalline structure at the atomistic level and the nano-sized polymer morphology.

A series of TPUs, produced from the same starting materials as the present TPU, with hard block contents (HBCs) ranging from 22 to 75 wt.%, were submitted to WAXS and SAXS. WAXS at room temperature showed that the crystallinity of the hard phase was only observed at HBCs above 35 wt.%, and even then, the degree of crystallinity remained in the single-digit range [3].

The hard domains at an HBC of 29 wt.% are the discontinuous phase surrounded by soft phase material. The hard domains are poorly arranged; there is no clear periodic arrangement, but some rudimentary quasi-periodicity exists. The uniformity in distance between the periodically arranged hard phases generates a scattering maximum, known as the SAXS long peak [Fig. 4.6a]. The average length of the hard-and-soft-domain repeating unit can be calculated from the scattering angle of the long peak using Bragg's equation. This repeating distance was calculated to be 10 nm. Detailed calculations revealed that the average thicknesses of the soft and hard layers were approximately 4 and 6 nm, respectively. During phase separation, the polyols act as tie molecules, binding neighboring hard domains and providing the required hard-soft interdomain distance. This uniformity, in turn, generates the SAXS long peak.

The temperature in the SAXS experiment was increased to 220 °C at the same rate as in the present DSC experiment. The SAXS long peak (yellow streak in Fig. 4.6b) remained constant in both position and intensity until 120 °C. However, it started shifting to smaller angles and losing intensity at increased temperatures. The peak shift to smaller angles indicates that the interdomain spacing increases; the hard domains become further apart. The gradual loss in intensity indicates a dissolution process of the

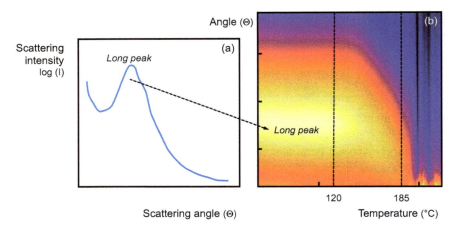

Fig. 4.6: The scattering intensity versus scattering angle showing the SAXS long peak (a) and the scattering intensity of the long peak (yellow streak) versus temperature (b) [3] (with the kind permission of John Wiley and Sons).

hard domains, which starts at 120 °C and is completed at about 185 °C. The dissolution process, determined with SAXS, matches the melting transition observed in the DSC experiment.

4.4.3 Softening of the hard domains

The G' trace of the present TPU, as shown in Fig. 4.3 (page 87), was combined with the DSC thermogram [Fig. 4.5]. The softening temperature of the TPU coincides with the onset of the melting and dissolution processes determined by DSC and SAXS, respectively.

The hard domains of the present TPU did not show crystallinity, and the melting transition by DSC corresponded to the dissolution process of the hard domains in SAXS. The observed endotherm in DSC indicates a process in which disorder increases as a result of the supplied energy required for the dissolution of the hard domains [4]. However, the mechanical stability of the PU already diminished at the onset of melting. When the supplied thermal energy is sufficient to start the dissolution process, the hard blocks in the hard domains gain thermal mobility. The hard domains soften, and the polymer begins to flow under strain. The melting is stretched over a long temperature interval because the hard domains in the TPU differ in size and strength. The latter may originate from the breadth of the hard block size distribution; longer blocks yield more stable hard domains that survive longer and delay the dissolution process [5].

4.5 Burning behavior and flame protection

Organic materials, such as polymers, burn when subjected to a flame, producing heat and smoke. Two basic models for burning have been proposed: the "fire triangle" and the stage model, which describes the various stages of a developing fire.

The fire triangle is the standard model for understanding the necessary conditions for combustion [Fig. 4.7].

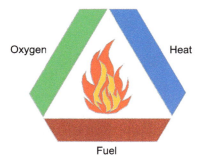

Fig. 4.7: The fire triangle: oxygen – heat – fuel.

A fire can only exist when all three components come together: oxygen, heat, and fuel. Oxygen is the fundamental component of combustion, and heat is required to degrade the polymer and generate combustible gases. Fuel is the source of fire, in the present case, the polymer. In most cases, the polymer does not burn; the combustible gases generated by the flame's heat fuel the fire. When one of the components is removed, the fire extinguishes. Covering a fire with a blanket (starving it of oxygen) or cooling it with water (removing heat) will extinguish the fire.

The process of combustion for organic materials has about six different stages [1]:
- The ignition source heats the material, raising its temperature.
- The heated material starts to degrade.
- The degradation speeds up, releasing low-molecular-mass decomposition products.
- With sufficient oxygen and an ignition source, the combustible gases ignite, and the combustion process starts.
- The gases combust at or near the surface of the organic material; if sufficient energy is produced, the process can become self-sustaining.
- The flames start propagating along the polymer surface, and charred surface layers may form, depending on the type of organic material.

A standard polyurethane polymer with an index of 100 without additives will completely burn. Fire resistance can be improved by increasing the polymer's intrinsic thermal stability or adding fire retardants.

Polyurethanes produced from aromatic isocyanates possess a relatively high aromatic content, which increases with increasing isocyanate content in the formulation.

The aromatic content can be further increased using aromatic polyester polyols. Furthermore, thermally stable crosslinks, such as isocyanurate, can be incorporated (PIR foam). The high aromatic content, combined with thermally strong crosslinks, enhances the polymer's char-forming tendency and its temperature resistance to thermal degradation [1]. The char resulting from sufficiently fire retarded rigid PU or PIR foam is a thermally stable aromatic carbonaceous structure. It acts as a gas diffusion barrier and shields the polymer beneath from thermal degradation. Rigid foams with acceptable flammability characteristics can be obtained by optimizing their aromatic and isocyanurate content [6].

4.5.1 Fire retardant mode of action

Various classes of fire retardants are available, each with a specific mechanism that interferes with the burning process. Flame retardants are generally incorporated by adding them to the A-component. Therefore, especially in foam applications, the fire retardants are preferably liquid so that they can be easily mixed into the polyol component and applied. Fire retardants are primarily used to prevent polymer ignition. Therefore, they should become active in the early stages of the fire development. The additives operate chemically or physically, depending on their chemical nature. The chemically active fire retardants used in PU are halogen- or phosphorus-based. The physically active additives, such as aluminum trihydrate and melamine, cool the flame by endothermic decomposition.

Phosphorus-containing flame retardants primarily act in the solid phase. Thermal degradation transforms the flame retardant into phosphoric acid, which catalyzes charring reactions in the polymer. The formation of the char protects the material beneath from further thermal degradation.

The mechanism of ammonium polyphosphate (APP) fire retardancy in PU was studied in detail. At high temperatures, a reaction occurs between APP and PU. This accelerates the polymer decomposition, but at the same time, it forms a high-temperature stable phosphocarbonaceous polyaromatic structure on the polymer surface. The high-temperature condensation reactions form the char, which eventually oxidizes at elevated temperatures or upon prolonged exposure to high temperatures. The char contained large radical species, such as polyaromatic macromolecules trapping free radicals, and was strongly paramagnetic [7].

The oxidation reactions proceed via free radical mechanisms involving hydrogen and hydroxyl radicals. Halogen-containing fire retardants yield hydrogen halides that interfere with the radical reactions, e.g., by capturing hydroxyl radicals, reducing the exotherm, and cooling the flame. The back radiation of the flame on the polymer is reduced, lowering the polymer degradation rate.

The activity of halogen and phosphorus fire retardants has been compared. Phosphorus is the most effective fire retarding element, followed by bromine, which, in turn, is a stronger fire-effective element than chlorine. The combination of phosphorus and halogen-based additives is claimed to exhibit synergistic effects [1].

Melamine is a fire retardant used in flexible foam. Upon heating, melamine undergoes endothermic degradation, forming gaseous products. Incorporating melamine into flexible foam effectively reduces the heat release rate and assists char formation, thus protecting the foam.

Inorganic compounds, such as magnesium hydroxide or aluminum oxide trihydrate, are used as fire retardants in compact PU materials, including TPU. In the case of fire, they are endothermically dehydrated to the oxides, and the liberated water dilutes the volatile gases in the flame.

Expandable graphite is used in foams with very low densities, such as automotive roofliner foam. The graphite flakes expand to a multiple (up to several 100 times) of their original volume upon heating. Three steps of the intumescence development were observed: formation, stabilization, and degradation of the intumescent shield. The thermo-oxidative degradation of the polymer and the char structure was not affected by the addition of expandable graphite. It was the formation of an intumescent layer on the material surface that reduced the spread of flame [8].

4.5.2 Flammability testing

Depending on the application, several tests are in place to evaluate the burning behavior of polyurethanes. Fundamentally, there are two classes of fire tests: resistance to fire and reaction to fire.

Fire resistance can be determined through destructive fire testing, designed to replicate the product's intended end-use, such as rigid foam panels for construction applications. The tests address fully developed fires and assess the impact of fire on the integrity of a panel assembly. The test is large-scale, up to the size of a garage, and the fire load is substantial, for instance, a gas burner or a crib of wood weighing several kilograms. Such tests are required for technical specifications set by regulations or insurance companies.

The small-scale tests are "reaction to fire tests," which address the initial stages of fire development, examining ignitability and fire propagation. In general, small pilot flames are used to ignite the material. The test results are used for research purposes, quality control, or compliance with industry standards. The tests commonly assess the burning behavior by characteristic values, such as ignition time, the amount of external heat required for ignition, the minimum amount of oxygen needed to sustain combustion, flame spread, or the heat and smoke released during burning. Three general tests are discussed for illustration. In the product sections, some application-specific flame tests will be presented.

UL 94

The UL 94 (UL = Underwriters Laboratory) test, harmonized with ISO 9772, is a general flammability test for plastic materials. It is used to test compact PU elastomers. A test bar of 125 mm × 13 mm and a thickness of 0.4 – 3 mm is exposed vertically to an open flame from below with a flame height of 20 mm (Fig. 4.8). The exposure time depends on the specific classification. The time till extinguishment and polymer dripping are recorded. To achieve the highest classification (V-0), the flame must extinguish within 10 s, and the drippings must not burn.

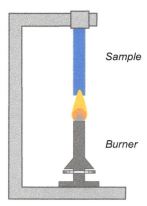

Fig. 4.8: Flammability test according to UL 94.

Limiting oxygen index

The polymer's limiting oxygen index (LOI, ISO 4589) is the minimum concentration of oxygen in the surrounding atmosphere, expressed as a percentage, required to sustain combustion. It is measured by passing a mixture of oxygen and nitrogen from below over a vertical specimen that is ignited at the top. The oxygen level is reduced until a critical level is reached – the minimum amount of oxygen needed to continue burning. With increasing LOI, the flame resistance of the polymer improves. The material becomes self-extinguishing when the LOI exceeds the atmospheric oxygen concentration of 21 vol.%; however, a material must be considered flammable as long as the LOI value is less than 26 vol.%.

Cone calorimeter

The cone calorimeter is a fire-testing instrument based on the principle that the amount of heat released from a burning sample is directly related to the amount of oxygen consumed during the combustion (ISO 5660). The total amount and the rate of heat a material releases may correlate with fire properties, such as the fire growth rate and flame spread. Concerning PU, it is predominantly used for rigid foam.

The material is exposed to an external radiant heat source. A sample of 100 mm × 100 mm and a thickness of approximately 50 mm is placed on a balance.

The sample is heated by a conical electric heater from the top, and sparks ignite the released gases. The combustion products travel through the cone heater into an instrumented exhaust pipe. The heat flux radiating to the sample can be varied. It can be held constant during the experiment at 35 or 50 kW/m^2 or increased stepwise, e.g., from 10 to 75 kW/m^2. The recorded values typically include the time to ignition, the mass-loss rate during combustion, the time to and level of maximum heat release rate, the total amount of heat released, and the rate of smoke production.

References

[1] D.W. van Krevelen, Properties of polymers, Elsevier, 4th Ed., 2009, ISBN 978-0-08-054819-7 (Chapters 5, 6, 13, 26).

[2] R.J. Young and P.A. Lovell, Introduction to polymers, CRC Press, 3rd Ed., 2011, ISBN 978-0-8493-3929-5 (Chapter 5).

[3] A. Stribeck, B. Eling, E. Pöselt, M. Malfois, E. Schander, Melting, solidification, and crystallization of a thermoplastic polyurethane as a function of hard segment content, Macromol. Chem. Phys., 2019, 220, 1900074.

[4] A. Stribeck, B. Eling, E. Pöselt, M. Malfois, E. Schander, Melting and solidification of thermoplastic polyurethanes as a function of nucleating agents, Nano Select, 2021, 1–15.

[5] Z. Gao, Z. Wang, Z. Liu, L. Fu, X. Li, B. Eling, E. Pöselt, E. Schander, Z. Wang, Hard block length distribution of thermoplastic polyurethane determined by polymerization-induced phase separation, Polymer, 256 (2022) 125236.

[6] A. Cunningham, B. Eling, D.J. Sparrow, A study of the low smoke potential, flame resistance, and processability of high-index polyisocyanurate rigid foam, J. Cell. Polym., 6 (6) 42–60 (1987).

[7] S. Duquesne, M. Le Bras, S. Bourbigot, R. Delobel, G. Camino, B. Eling, C. Lindsay, T. Roels, H. Vezin, Mechanism of fire retardancy of polyurethanes using ammonium polyphosphate, J. Appl. Polym. Sci., 82, (13), 3262–3274 (2001).

[8] S. Duquesne, M. Le Bras, S. Bourbigot, R. Delobel, G. Camino, B. Eling, C.I. Lindsay, T. Roels, Thermal degradation of polyurethane and polyurethane/expandable graphite coatings, Polym. Degr. Stab., 74, 493–499 (2001).

5 Processing

Polymer processing involves converting a polymer into the desired shape or form. Thermoplastic polymers are physically transformed by molding, extrusion, or calendaring. In contrast, PU articles are manufactured by converting liquid components into a polymer. The reactants are machine-mixed, and the polymerization reaction is carried out continuously or discontinuously.

The polyol blend and isocyanate are metered in the desired ratio into a high or low-pressure mixing head. After mixing, the reactive liquid is discharged into a mold or on a belt, where it polymerizes. Molding is a discontinuous process of producing PU articles that are ready for further use. In a continuous process, the reactive mixture is conveyed on a belt into a heated tunnel for curing. After the reaction is completed, the bun or sheet is cut into slabs or panels for further use. Some general principles of the different PU production technologies are given.

5.1 Prepolymer process and one-shot method

The prepolymer process was predominant in the early days of industrial PU manufacturing. First, the polyol was reacted with an excess of polyisocyanate to form an NCO prepolymer. Subsequently, the prepolymer was chain-extended with short-chain glycols or water to form a compact or foamed PU. Today, this process is only used in some specialty applications, such as high-end elastomers.

The invention of urethane-specific catalysts significantly altered the manufacturing process of polyurethane. Since then, PU production has used one-shot or semi-prepolymer technology. The former uses a basic polyisocyanate, whereas the latter uses an isocyanate prepolymer in which part of the polyol is pre-reacted with the isocyanate. The polyol blend, containing all the required additives, and the isocyanate are metered into a mixing head, mixed, and processed continuously or discontinuously.

5.2 Discontinuous and continuous processing

Discontinuous processing involves the following steps: calculating the A-to-B ratio, setting up the mixing machine, preparing the mold, filling the mold, demolding, and post-treatment.

The formulator develops the formulation and selects the appropriate index. This gives the mass ratio of the A- and B-component. The machine settings can then be calculated from this ratio and the weight of the liquid mixture required to fill the mold.

https://doi.org/10.1515/9783110744583-005

The surface quality of the molded PU foam is determined by the amount of reactive mixture used to fill the mold. In general, the amount of reactive mixture injected is higher than theoretically required, resulting in an overpacking of the mold. The theoretically required amount of PU is calculated from the "free rise" density of the foam and the volume of the mold. The foam produced in a cup, bucket, or bag (called the "bag shot") determines the free-rise density. The overpack factor can vary significantly and depends on the system and application. For example, this ratio typically ranges from 1.1 to 1.3 for a molded PU flexible foam.

PU reaction mixtures show strong adhesion to substrates. Therefore, the mold surfaces have to be protected against adhesion to PU. In plain molds, such as a block mold, this can be accomplished by protecting the surface with foil or inserting a plastic bag. More complex molds require external mold release agents, which are typically applied by spraying the release agent onto the mold's surface. Self-releasing PU formulations can be developed by incorporating an internal mold release agent to the polyol component.

After mold preparation, the required amount of reaction mixture is dispensed into the mold [Fig. 5.1]. Systems with a relatively slow reactivity can use open molds, which are closed after filling. More reactive systems generally use closed mold injection. Important are the times it takes to fill the mold – the fill time – and the time the foam starts expanding – the cream or start time. The fill time has to be shorter than the cream time. This prevents the already foaming material from coming into contact with the fresh liquid material because this would inevitably lead to foam defects where the two fronts meet.

Fig. 5.1: Foam injection into an open mold (with the kind permission of Hennecke GmbH).

The reactive liquid expands and fills the mold. When the polymer is strong enough, the mold can be opened, and the PU article demolded – the demolding or handling time. After demolding, the article typically requires post-treatment, which may involve handling steps such as cleaning, matting, and trimming. Finishing the part may involve coating or adhering it to other materials.

The primary applications for continuous processing are the production of flexible slab stock and laminated rigid insulation boards. The essence of both production processes is that the reactive mixture is continuously discharged on a conveyor belt. The foam starts expanding, and when it has reached its final height, it passes a heated tunnel, called the conveyor, to accelerate the curing process. After attaining sufficient mechanical strength for handling, it is cut into smaller parts, stacked, and stored for further use. Flexible foam production yields large foam blocks for use in mattress and furniture manufacturing, whereas the rigid foam process yields insulation boards and panels ready for immediate use. The processing machinery for continuous foam production is very application-specific and will be discussed in Chapters 7 (rigid foams) and 8 (flexible foams).

5.3 Low-pressure processing

Low-pressure mixing heads consist of a hollow cylinder with a high-intensity mechanical stirrer. In the standby position, the polyol and isocyanate components are individually recirculated at pressures ranging from 3 to 40 bar from the feed tank through the mixing head and back. The metering pumps for recirculation are generally gear pumps. During the release of a "shot," the valves open, and the monomers are injected into the mixing chamber and stirred. The stirrer generates high shear forces required for forming tiny air bubbles, essential for foaming. The shot time and the reactant flow rates determine the amount of material injected into the mold. After the shot is completed, the valves are closed, and the polyol and isocyanate are recirculated again [Fig. 5.2].

Optionally, additional components, such as additives, color pastes, and blowing agents, can be metered directly into the mixing head. After each shot, the mixing chamber must be cleaned to prevent clogging by the residual reaction product left behind. The cleaning is accomplished by flushing a cleansing solvent through the mixing chamber. Cleaning after each shot takes time and generates waste.

The output rates are low, ranging from 0.3 to 10 L/min, and the mixing times are relatively long, lasting up to several seconds. The long mixing times ensure excellent mixing but limit low-pressure processing to systems with slow reactivities. It has the advantage that highly viscous components (up to 40,000 mPa·s) and systems with mixing ratios that deviate significantly from a 1:1 ratio can be processed. Low-pressure technology is preferred for producing small parts such as shoe soles.

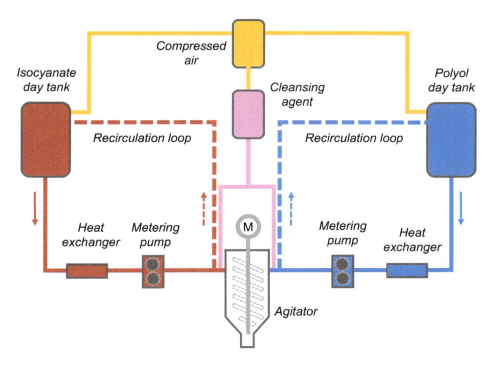

Fig. 5.2: Schematic illustration of the low-pressure mixing process.

5.4 High-pressure processing

High-pressure machines, especially those used for discontinuous processing with short shot times, operate in a recirculation mode of the reactants. The reactants are recirculated at pressures of about 100 – 200 bar.

In the standby position, the components are circulated from the day tank to the mixing head and back via grooves in the control piston [Fig. 5.3]. The most common types of metering pumps are plunger and axial piston pumps, although gear pumps are also used. The shot begins by retracting the control piston, creating a mixing chamber with volumes of 0.1 – 10 mL. The polyol and isocyanate are injected from opposite directions via narrow nozzles at high pressures. The components impinge at high speeds, and mixing occurs within milliseconds ("impingement mixing"). After the shot, the control piston is reversed to standby, closing the mixing chamber and cleaning the shaft. The mixing head is self-cleaning because of the precise fit between the piston and shaft.

To ensure efficient mixing, the viscosity of the components should be relatively low, preferably below 5,000 mPa·s, and the A-to-B volume ratios should not deviate significantly from parity. Multi-component mixing heads can handle additional

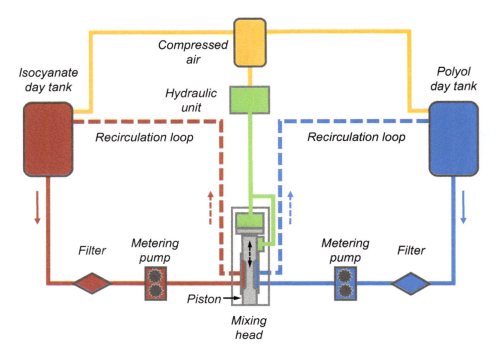

Fig. 5.3: Schematic drawing of the high-pressure mixing process with recirculation.

components, such as blowing agents, as a separate stream. With an output rate of 2 – 400 L/min, high-pressure processing is well-suited for producing large articles, such as car body panels, and for continuous processing. Because the mixing times are extremely short, systems with high reactivities can be handled.

Most high-pressure dispensers operate in a recirculation mode; however, some dispensers may use the older "inline" technology. The inline technology directly feeds the components from the metering units to the mixing head. The nozzles for component injection are closed by hydraulically preloaded springs and open as soon as the pressure build-up is high enough to overcome the preload of the holding spring. This technology can be utilized for continuous processing, such as in the production of rigid foam laminated boards. At the end of the production, the pressure is rapidly reduced via pressure relief valves, thereby closing the pressure-controlled nozzles.

5.5 RIM technology

Reaction Injection Molding (RIM) is used for the manufacturing of high-density and compact parts for the automotive industry with densities of approximately 1,000 kg/m^3. The materials can be soft and elastic, such as sealings for window encapsulation, or hard and tough, such as car body panels. RIM machines allow the processing of highly reactive systems such as amine-containing A-components for polyurea RIM.

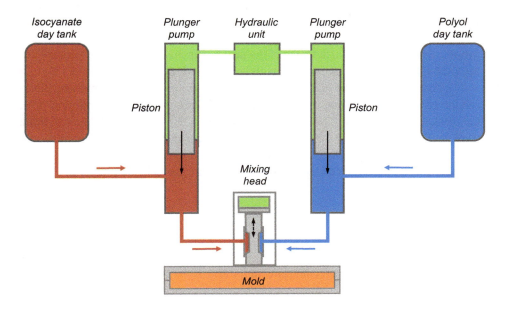

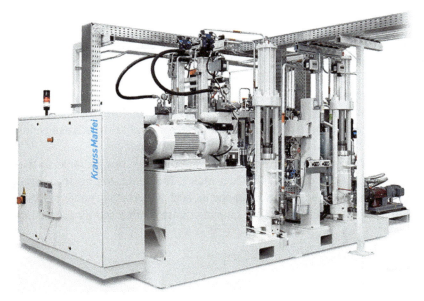

Fig. 5.4: Top: Flow chart of the RIM process; bottom: high-pressure metering machine with tandem plunger pumps (with the kind permission of KraussMaffei Technologies GmbH).

Most RIM machines utilize tandem piston or plunger pumps for injection [Fig. 5.4]. The A- and B-components are metered to the cylindrical chambers at low pressures. During the stroke, the pressure in the cylinders increases to the set processing pressure, at which the pressure-controlled injection nozzles open, starting the shot. The plungers are usually driven hydraulically. The size of the two cylinders allows for the discharging of the reactants in a single stroke of the piston. The machine can handle highly viscous and filler-containing components with flow rates of up to 10 kg/s.

Standard high-pressure RIM machines are generally unsuitable for producing fiber-reinforced composite systems. Special technologies, such as Long Fiber Injection (LFI), Composite Spray Molding (CSM), and Reinforced Reaction Injection Molding (R-RIM), were specifically developed for this purpose. Glass fiber mat-reinforced composites can be manufactured using Resin Transfer Molding (RTM) or Structural RIM (S-RIM). Profiles containing continuous glass fibers can be produced using the pultrusion process.

5.6 Equipment

Large-scale polyurethane processing requires facilities for storing and handling raw materials, molding equipment, metering units, and dispensers.

5.6.1 Storage

Depending on demand, polyisocyanate and polyol are delivered in drums (200 – 250 kg), Intermediate Bulk Containers (IBCs; ≈ 1.2 t), or in tank trucks (≈ 23 t). In large plants, raw materials are stored in tank farms, typically featuring pairs of storage tanks for isocyanate and polyol, with volumes ranging from approximately 15 to 30 m^3. The starting components are stored in closed systems to protect them against humidity and temperature fluctuations. Special attention has to be paid to the storage conditions of the isocyanate to avoid crystallization and solids formation.

The starting materials are transferred from the large storage tanks to the significantly smaller working or day tanks ("day tank"). The liquids are pumped through premixing stations, where additional components, such as surfactants and catalysts, can be added. In the day tanks, the reaction components are conditioned for processing.

5.6.2 Molds

Molds have to fulfill several requirements and need the following functions [1]:
- Receive and distribute the reaction mixture.
- Maintain temperature.
- Accommodate the pressure.

- Prevent material loss and reduce flash.
- Avoid air entrapment.
- Fixation of inserts if required.

The mold temperature can be critical for the quality of the final article. Cold molds retard surface cure, which may lead to the formation of brittle and dense skins and poor adhesion to substrates. Therefore, molds are usually warmed up before the campaign begins. During the day, however, the molds heat up because of the exothermic reactions. When the mold surface temperature becomes too high, the blowing agent may evaporate prematurely, resulting in the formation of subsurface voids and blowholes. Therefore, molds often contain cooling units to dissipate the heat.

The pressure in a closed mold increases significantly and, depending on the overpacking, may reach 100 kPa or more. Therefore, the molds and mold carriers are typically designed to withstand 200 kPa, corresponding to a clamping force of approximately 20 tons for a medium-sized mold of $1\,m^2$.

During the molding process, entrapped air must be able to escape. This is accomplished via venting holes, usually located at the top or along the parting lines of the mold.

Molded PU flexible and microcellular foams may need stiffeners for applications such as steering wheels and headrests. The insert can be a metal frame or a molded thermoplastic part. The molds are then equipped with attachment points to hold the insert in place during the molding of the PU foam.

In-mold decoration
PU parts can be coated after production as a post-treatment step. This step can be omitted using In-Mold Coating (IMC) technology. A coating is sprayed onto the mold surface before the reactive liquid is discharged. Upon demolding, the coating is transferred from the mold to the PU. Hence, the coating acts as both a mold-release agent and a decorative element for the PU article.

Dashboards consist of semi-rigid foam backed by a rigid insert and covered by a decorative skin. The rigid insert is typically fixed to the upper part of the mold, whereas the preformed skin is placed and held in position at the lower part of the mold. The mold is closed, and the PU foam is injected into the cavity between the two substrates. This is closed mold injection; however, open mold injection technology is also applied. After sufficient cure, the dashboard can be demolded. This technology is also referred to as pour-behind technology.

Multiple mold processing
To enhance productivity and reduce cost, the molds are arranged such that a single dispenser can serve several molds. These molds can be attached to a fixed or rotating table. Fixed tables are primarily used with open molds, where the mixing head is positioned above the mold and then follows a pour pattern, filling one mold after

another. Rotary tables can accommodate more than 20 molds. The mixing head is stationary, and the molds are moved to and from it [Fig. 5.5]. Open-pour or closed-mold injection may be applied. Fixed and rotary tables are used, for example, in the production of automotive seating and shoe soles, respectively.

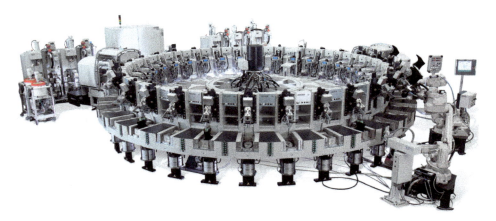

Fig. 5.5: Rotary table with open molds for shoe sole production (direct soling, with the kind permission of DESMA Schuhmaschinen GmbH).

5.6.3 Metering units

The most commonly used metering units are gear, plunger, piston, and axial piston pumps.

Gear pumps
Gear pumps can be run at various pressures and handle highly viscous and filler-containing components. At the inlet side of a gear pump, the liquid is trapped and transported in the pockets between the teeth of two rotating gears, and then released at the outlet side of the pump. The gear drive speed controls the output.

Plunger and piston pumps
In plunger and piston pumps, the cylindrical chamber's working volume is reduced and enlarged by the plunger or piston, which is eccentrically attached to a rotor, causing it to move up and down upon rotation. During the suction stroke, the volume in the cylinder increases, and the liquid is sucked in via a one-way valve. The volume is reduced via a second one-way valve during the compression stroke, pressing the liquid in the pumping direction. This pumping system is disadvantaged because the liquid does not move uniformly but periodically with each suction-compression cycle.

The primary difference between a plunger and a piston pump lies in the technique used for sealing the cylinder. In a plunger pump, the sealing is stationary at one end of the cylinder, and the plunger moves through the sealing at each stroke. In contrast, in a piston pump, the sealing is attached to the piston, which moves within the cylinder. Plunger pumps can handle pressures of up to 2,000 bar.

Axial piston pumps
The periodic pumping of a single piston is evened out in axial piston pumps. These pumps contain several, typically seven, pistons [Fig. 5.6]. The pistons are arranged in a circular pattern within the rotating cylinder block. The pistons are attached to a fixed plate positioned at an angle to the cylinder block, the cam angle. Due to the angular geometry, the pistons now move sinusoidally as the cylinder block rotates. One end of the cylinder block is sealed with a valve plate that consists of two ports, one for the component inlet and one for the outlet. These ports are arranged so that the pistons connect to the inlet port during the suction and outlet port during the compression stroke. The flow rate is changed by adjusting the cam angle, which determines the volume change in each piston during a stroke.

Fig. 5.6: Axial piston pump (left: with the kind permission of Bucher Hydraulics GmbH; right: with the kind permission of Hennecke GmbH).

5.6.4 Mixing heads

Industrial foam production requires machine mixing of the reactants. The mixing is performed with either a low-pressure or a high-pressure mixing head.

Low-pressure mixing heads

The standard low-pressure mixing heads are agitator mixing heads. Polyol and isocyanate are supplied via input injectors located at the opposite sides of the mixing chamber. The rotation of the stirrer provides the mixing [Fig. 5.7]. Its rotational speed ranges from 1,000 to 20,000 rpm. The stirrers exist in different forms and shapes, such as spikes, paddles, screws, or propellers. These mixing heads can handle highly viscous and filled components. The required cleaning of the mixing chamber after each shot is a disadvantage of this mixer type.

Fig. 5.7: Examples of agitator mixing heads (left: with the kind permission of Hennecke GmbH; right: with the kind permission of Cannon S.p.A.).

High-pressure mixing heads

High-pressure mixing heads are self-cleaning. Several high-pressure mixers exist, but all work according to the same principle. Fig. 5.8 illustrates an example of a linear mixing head.

The hydrolytically driven control piston separates the isocyanate from the polyol and switches the mixing head between recirculating and mixing. When the control piston is in position (a), the isocyanate (red) and polyol (blue) are recirculated via the longitudinal grooves in the piston. After lifting the piston (b), the components are injected into the mixing chamber under high pressure and mixed. The control piston is pushed down to stop the shot and, at the same time, closes and cleans the mixing chamber (c).

The injection speed of the components into the mixing chamber through the small orifices is high, reaching up to 150 m/s, and the flow is turbulent. The turbulent flow ensures good mixing; however, for a proper laydown of the reactive liquid in the

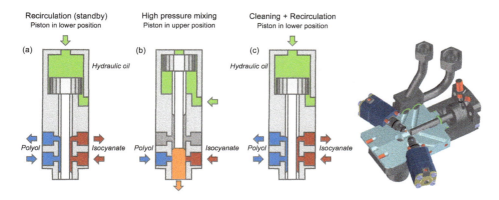

Fig. 5.8: Linear high-pressure mixing head: (a) recirculation; (b) mixing; (c) cleaning (with the kind permission of Hennecke GmbH).

mold or on the conveyor belt, the reactive liquid ideally needs to leave the nozzle as a laminar flow. For this, the flow rate must be reduced significantly, to below 3 m/s. This can be accomplished, for instance, with an L-shaped outlet at right angles between the orifice of the mixing chamber and the outlet pipe [Fig. 5.9]. This geometry establishes laminar flow by perpendicular deflection and increased outlet pipe diameter.

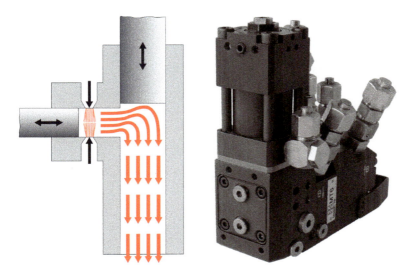

Fig. 5.9: L-shaped mixing head (with the kind permission of Hennecke GmbH).

References

Further reading

- W. Becker, Reaction Injection Molding, Van Nostrand Reinhold Company, New York, 1979, ISBN 978-04-42206-31-4.
- U. Knipp, Herstellung von Großteilen aus PUR-Schaumstoffen, Zechner & Hüthig Verlag, Speyer, 1974, 72–77.
- G. Oertel, Polyurethane, Kunststoff Handbuch (Becker/Braun), Band 7, 1993, ISBN 3-446-16263-1, Chapter 4, 139–192.

Special references

[1] R. Leppkes, Polyurethanes, a versatile specialty plastic, 2012, ISBN 978-3-86236-039-0, 23–30.

6 Foam formation

The main application of PU is in foams, notably rigid and flexible foam. Foaming requires chemical and/or physical blowing agents. The most used chemical blowing agent is water. Its reaction with isocyanate gives carbon dioxide, which is used to expand the foam. Physical blowing agents are low-boiling liquids that evaporate due to the heat generated by the chemical reactions.

When the isocyanate and polyol blend are mixed, tiny gas bubbles nucleate the reaction mixture. The tiny bubbles initiate the growth of the individual cells and are essential for developing the cellular structure. Once the chemical reactions start, gases flow from the liquid into the bubbles, and because the gases are trapped in the cells, the foam expands.

Foams are generally produced at densities of 30 – 50 kg/m^3, which means that the foam volume during expansion increases by a factor of 20 – 30. The growing cells need sufficient stability during foam expansion to withstand the foaming process. Cell stability can be improved physicochemically by adding a surfactant and chemically increasing the reaction rate. With time, the reactive mixture polymerizes, the mechanical strength develops, and the expansion stops. The foam has then attained its final density and morphology [Fig. 6.1].

Fig. 6.1: The different stages in the foaming process: (1) metering of polyol and isocyanate; (2) mixing and formation of nuclei; (3) creaming and initial rise; (4) foam expansion, bubble packing, and polymer gelling; (5) solidification (with the kind permission of BASF SE).

Open- and closed-cell foam can be produced [Fig. 6.2]. The cells in rigid foam are closed, whereas flexible foam is open-cell. The closed-cell nature of rigid foams is fundamental to their good heat insulation properties. The open-cell structure of flexible foam allows for intercellular airflow upon deformation. High air permeability is a basic requirement for their cushioning properties. The end properties of the foam depend on the foam density, the structure and geometry of the cells, and the properties of the polymer matrix material.

https://doi.org/10.1515/9783110744583-006

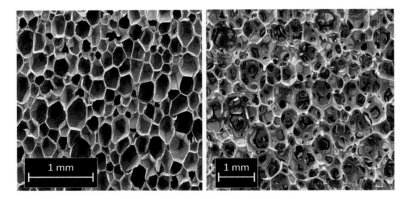

Fig. 6.2: SEM-pictures of PU foams: (left) closed-cell rigid foam; (right) open-cell flexible foam (with the kind permission of Covestro AG).

6.1 Simultaneous formation of polymer and foam

Foam formation from liquid PU monomers is complex because foam expansion and the buildup of the polymer structure occur simultaneously. To explain the formation of flexible and rigid foam, the polymerization reaction and the foaming process are conceptually separated.

6.1.1 Flexible foam formation

The primary reactants for producing a flexible foam are polyol, isocyanate, and water, and the main additives are catalyst and surfactant. Water is used as the chemical blowing agent. Its molar amount is significantly higher than that of the hydroxyl groups of the polyol, so the number of urea linkages in the foam is significantly higher than that of urethane. The water content is only a few percent of the polyol by weight. Considering the mass amounts of polyol and isocyanate, the polyol is always the larger component. The polyol, water, and additives are mixed to give the polyol blend. Subsequently, the polyol blend and isocyanate are mixed. These two components may be miscible, but they are often incompatible. This is, for instance, the case when mixing a polyether polyol blend with a polymeric MDI-based isocyanate.

Intensive mixing of the two components yields an emulsion of isocyanate-in-polyol with droplet dimensions ranging from 1 to 10 µm. During the early stages of the reaction, block copolymers of the A-B type are formed, where A represents the polyol moiety and B represents the isocyanate moiety. These block co-polymers act as compatibilizers, reducing the interfacial tension between the two phases. The homogenization process is further enhanced by bubble expansion, which causes liquid flow at the intersections between bubbles. A homogeneous polymerizing mixture is generally, but not necessarily, obtained. The reaction proceeds, and more urethane and urea linkages are formed. A reaction-induced phase separation occurs when the urea hard blocks have surpassed a critical segment length. As a result of the phase separation, a nanometer-sized polymer morphology consisting of hard and soft phases is obtained. The polymerization process continues, the polymer first gels and then hardens, resulting in an elastic PU polymer.

Parallel to the polymerization reaction, foam formation occurs. During mixing, the reactive liquid is nucleated with tiny air bubbles. On a molecular level, the compatibilization of the monomers and the polymerization proceed, and the carbon dioxide formed in the urea reaction diffuses into the nuclei when the reactive mixture is saturated with carbon dioxide. Macroscopically, the foam starts to expand. The increase in molar mass and viscosity gives stability to the expanding foam. The internal and external gas pressure difference drives the expansion of the foam. The gelation process triggers cell opening [Chapter 8.4.1]. The foam expansion stops upon cell opening, and the blowing agent escapes. The continuing reaction hardens the polymer, resulting in a dimensionally stable open-cell foam.

6.1.2 Rigid foam formation

When comparing the foam expansion of rigid foam with that of flexible foam, several commonalities and fundamental differences are evident. Flexible foams use only water to expand, whereas a combination of water and a physical blowing agent is typically used in rigid foams. Their preparation methods, expansion rates, final foam densities, and cell sizes are comparable but differ in polymer and foam morphology. Rigid foams are produced from PMDI and polyols with low equivalent masses and high functionality. The amounts of polyol and isocyanate are approximately equal, and the water content in the polyol formulation is about half that of flexible foam. In contrast to flexible foam, the urethane content in rigid PU foam is significantly higher than that of urea. Using branched monomers results in a rapid buildup of molar mass and early chemical gelation, thereby stabilizing the cells during expansion. The high crosslinking prevents phase separation, resulting in the formation of an amorphous and rigid polymer. Cell opening does not occur, and the expansion stops when the polymer has solidified. The mechanical strength properties build up quickly, resulting in the formation of closed-cell rigid foam.

6.2 Aeration and nucleation

The initial stage of foam growth is controlled by two bubble-forming mechanisms: aeration and nucleation [1]. Aeration refers to the entrainment of air in the reactive mixture, whereas nucleation describes the process of bubble formation within the liquid. Bubble formation requires energy, which is provided as part of the mixing energy. Which bubble-forming mechanism takes place depends on the applied mixing method. The reactive mixture is aerated when preparing a hand-mix cup foam, as shown in Fig. 6.1 (page 111). Bubble formation in high-pressure mixing occurs via nucleation, whereas both mechanisms can occur in low-pressure mixing. Foam producers do not necessarily distinguish between aeration and nucleation and generally refer to this initial bubble-forming process as nucleation.

Aeration is the primary mechanism for producing bubbles in cup foam. During vigorous mixing of the two components using a propeller blade mixer, air is whipped into the liquid, generating the initial small-bubble froth.

Nucleation is a fundamentally different process and, in principle, competes with aeration. It is a process where tiny gas bubbles are generated from gases dissolved in the reactive mixture. Because a bubble in a liquid has a much higher energy level than an equal number of gas molecules in solution, the system must exceed the surface energy barrier to produce that bubble as a new phase. The gases that form the initial bubbles, notably air and carbon dioxide, are available in significant amounts in the polyol and isocyanate. The gases were entrained into the starting components during their production or storage. For example, the amount of dissolved nitrogen in polyol at atmospheric pressure can be as high as 0.1 L per 1 L of polyol [2], whereas for carbon dioxide in isocyanate, this can be up to 1 L per 1 L of isocyanate.

Nucleation can occur from supersaturated gas/liquid solutions. This implies that conditions must prevail during the mixing stage at which the reactant mixture is supersaturated. The driving force for nucleation is the energy gain liberated during the transition from a single-supersaturated phase to a two-phase system consisting of a saturated liquid solution and gas bubbles.

Industrial foam making requires machine mixing of the polyol and isocyanate. Two different types of mixing heads are used in the industry: low-pressure and high-pressure.

Low-pressure mixing heads consist of a cylinder equipped with a mechanical stirrer. The two components are metered into the mixing chamber, mechanically stirred with high intensity, and then discharged. The cylinder has an air bleed, generally positioned at the top and opposite the discharge nozzle. Small amounts of air are injected into the mixing head and finely dispersed into the reactant stream during mixing. Aeration alone may already be sufficient for the foaming process. However, nucleation may also take place during low-pressure mixing. The rotational speed of the mixer is high, and locally, high shear forces are generated, especially at the trailing edges of the mixing blades. The energy input is generally high enough to allow the formation of nuclei by the dissolved gases in the reactants.

The mixing chamber of a high-pressure mixing head is significantly smaller than that of a low-pressure mixer, and the reactive components are subjected to high pressure, typically in the range of 100 – 200 bar. When the shot is made, the reactants impinge, mix turbulently, and are discharged from the mixing chamber into the open atmosphere. The energy conversion during countercurrent injection under high pressure and subsequent pressure release is utilized for the work required to mix the two components and nucleate the reactant stream. The nucleation mechanism in a high-pressure mixing head is similar to what happens when opening a soda bottle under pressure: the dissolved gases generate fine bubbles due to the sudden pressure drop. Because of the closed mixing head geometry, aeration is not possible. Thus, nucleation is the sole mechanism of bubble formation in high-pressure mixing [2].

6.3 Bubble growth

During the foaming process, bubbles are destroyed. The most substantial reduction in bubbles occurs straight after mixing when the bubbles are still very small and the viscosity of the reaction mixture is low. The bubbles are initially round and isolated in the reactive mixture, but they begin to touch each other upon growth, a phenomenon known as jamming. With further expansion, the foam structure develops.

In this context, foams at densities below the density at which jamming occurs are referred to as "bubbly liquids." Somewhat above this density, the foams are classified as "wet" or high-density foams. Towards the end of the foam expansion process, at densities below 100 kg/m^3, the foams are referred to as "dry" or low-density foams. Below 50 kg/m^3 densities, the cells deform and attain a polyhedral cell structure.

Surfactants are necessary for producing low-density foam. Surfactants stabilize the thinning bubble membranes during foam expansion.

6.3.1 Initial bubble stability and growth

The spontaneous formation of bubbles, or self-nucleation, is highly unlikely because the energy barrier for creating bubbles is far too high [1, 2]. Foam growth begins with the initially formed bubbles during mixing. These bubbles expand because gas diffuses from the liquid phase into the bubbles. Fig. 6.3 depicts a schematic drawing of a bubble with radius r in the reactive liquid. The surface tension of the liquid (γ) is on the order of $3 \cdot 10^{-2}$ N/m, and the initially formed bubbles are micron-sized. The internal pressures in the small cells are high, on the order of several bar. The initially formed bubbly liquid is highly unstable, and the bubble decay rate is high.

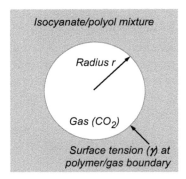

Fig. 6.3: Bubble with radius *r* in the reaction medium.

The tiny bubbles in the foam are metastable and, depending on their size, may re-dissolve or grow. According to the theories of Young and Laplace, the critical bubble radius (r_c) at which the bubble will be stable in the liquid requires a gas pressure differential given by:

$$P_b - P_l = \Delta P = \frac{2\gamma}{r_c}$$

P_b and P_l are the gas bubble and liquid phase pressures, respectively. The value of r_c for PU systems is estimated to be 0.25 µm [3].

For dispersing a gas in a liquid, the increase in free energy (*F*) equals the liquid surface tension multiplied by the interfacial area of the bubbles (*A*). Hence

$$\Delta F = \gamma \cdot A$$

The effect of the surface tension of the liquid on the foaming process can thus be expressed as:

$$\gamma = \Delta P \cdot \frac{r_c}{2} = \frac{\Delta F}{A}$$

This means that reduced surface tension will facilitate the formation of bubbles, as it reduces the energy required to form the bubble and favors the formation of smaller bubbles [2].

The air bubbles in the freshly generated froth are spherical but heterogeneous in size, ranging from 1 to 100 µm. The air volume fraction in machine-mixed foams is approximately 10 – 15%, and the number of air bubbles per unit volume is approximately 10^6 bubbles/cm³ [1]. The number of smaller bubbles decreases over time, while the larger bubbles increase in size. The theories of Young and Laplace can be used to describe this fundamental process of bubble loss. The pressure difference between the gas in a small bubble and that of an adjacent larger bubble is given by:

$$\Delta P = P_1 - P_2 = 2\gamma \cdot \left(\frac{1}{r_1} - \frac{1}{r_2}\right)$$

In this equation, P_1 and P_2 represent the gas pressures, and r_1 and r_2 represent the radii of the small and large bubble, respectively. When the radius of the big bubble is much larger than that of the small one, the term $1/r_2$ can be neglected, and the pressure difference causing gas diffusion is then proportional to the pressure in the small bubble. The rate at which small bubbles disappear depends on their radius, the inter-bubble distance, and the properties of the liquid that separates the bubbles. Bubble loss occurs during foam expansion as long as the polymer is liquid, and it stops when the polymer has sufficient strength to stabilize the cell structure.

6.3.2 Coalescence and disproportionation

The three main physical processes contributing to bubble loss and foam coarsening are coalescence, disproportionation, and drainage. These processes are interrelated and can occur simultaneously during foaming.

Coalescence is the process whereby two bubbles merge to form a single daughter bubble. The fusion results from the rupture of the thin liquid film formed between the bubbles upon contact. Disproportionation, also known as Ostwald ripening, refers to the inter-bubble gas diffusion because of gas pressure differences between the bubbles. Due to the high pressure, cell gas diffuses from the bubble into the surrounding liquid, and as a result, the gas concentration in the liquid phase surrounding the bubble increases. This leads to a concentration gradient of dissolved gas from small to large bubbles. Accordingly, gas starts to diffuse into the larger gas bubbles at the expense of the smaller bubbles. The smaller bubbles shrink and eventually disappear [Fig. 6.4].

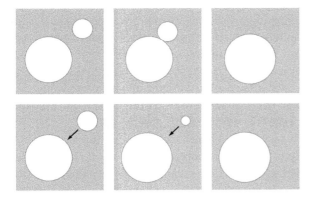

Fig. 6.4: Representation of bubble loss by coalescence (top) and disproportionation (bottom). The black arrow represents the gas diffusion process through the liquid.

6.3.3 Bubble jamming

The first cell membranes form when the expanding foam reaches densities of approximately 300 kg/m^3. With further foam expansion to densities of 200 kg/m^3, more membranes per cell are formed [Fig. 6.5]. This is the density regime at which the bubbles pack and form a bubble network. The onset of bubble packing is also referred to as the jamming transition. For a standard low-density foam with a density of approximately 35 kg/m^3, the bubble network forms when it has reached approximately 10 – 20% of its final height; hence, bubble packing occurs relatively early during foam expansion.

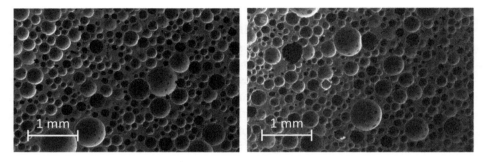

Fig. 6.5: Scanning electron microscopy (SEM) pictures of a rigid foam at a density of 290 kg/m^3 (left) and 210 kg/m^3 (right). The scale bar corresponds to 1 mm. Note the presence of the cell membranes and that the cell size distribution is broad (with the kind permission of BASF SE).

At the jamming transition, the foam's rheological properties undergo drastic changes [4]. From the start of mixing till the bubble network forms, the foam is a suspension of isolated gas bubbles in a liquid, and its viscosity, in principle, decreases with increasing gas content. When the bubble network forms, the foam attains solid or gel-like properties, similar to shaving cream, resulting from the interaction of closely packed bubbles. The bubble network still has a low modulus and expands easily when more gas diffuses into the cells. After bubble jamming, the foam grows steadily because of continuous gas diffusion from the liquid into the cells. The foam temperature increases due to exothermic reactions, further expanding the cells.

The onset of bubble network formation is noted during mold filling. When the low-viscosity reactive mixture is poured into the mold, it spreads and flows under the influence of gravity. The foam viscosity suddenly increases when the bubble network forms, and liquid spreading stops. From then onwards, mold filling occurs through foam expansion.

6.3.4 Drainage and foam structure development

Drainage refers to liquid flow due to gravity, capillary forces, and foam expansion. Gravitational forces can cause liquid drainage, but in the production of PU foam, this process is largely suppressed by the relatively high initial liquid viscosity, which increases rapidly due to polymerization. Capillary forces come into play when the bubbles begin to touch, and cell membranes form. The capillary forces cause suction and thinning of the membranes, which is further amplified by the expansion of the foam. The liquid polymer further concentrates in the intersections of the gas bubbles, and the foam becomes drier. The channels and junctions between bubbles are known as Plateau borders and nodes.

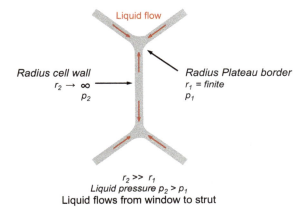

Fig. 6.6: The intersection perpendicular to two neighboring Plateau borders and their common window.

The gas pressure inside the individual bubbles is the same and, at this stage, approximately equal to atmospheric pressure. Fig. 6.6 illustrates the intersection of two neighboring Plateau borders and their shared membrane. The curvature radius of the Plateau border (r_1) has a finite value, whereas that of the cell membrane (r_2) goes essentially to infinity. The pressure within the planar membrane equals that of the cell, whereas the pressure in the Plateau border region is lower. This creates a suction force for liquid drainage from the membrane into the Plateau borders.

The spherical foam distorts into a polyhedral structure during the second half of the foam expansion. With continued expansion, the cell size increases, the membranes and Plateau borders become thinner, and the foam structure becomes more filigree. The continuous thinning of the membranes causes them to lose mechanical strength, and rupture may occur because of thermal or other mechanical disturbances during foam rise. Another possible reason for mechanical instability is the presence of small incompatible particles or droplets.

The chemical blowing reaction forms urea, which may form insoluble, microscopically small solid particles. As long as these particles are smaller than the membrane thickness [Fig. 6.7, left], they are enclosed by the liquid reaction mixture, and the membrane is stable. However, with further thinning of the membrane, it may become thinner than the particle. Depending on the surface characteristics of the particle, the membrane can then withdraw from the particle and rupture [Fig. 6.7, right]. This mechanism of cell opening is not limited to urea particles and rigid foam. Incompatible oils, deliberately added or present as contaminants, can also trigger cell opening. This bubble loss mechanism can occur in rigid as well as flexible foam.

Fig. 6.7: Cell opening in a steadily thinning membrane caused by incompatible particles in the membrane.

Fundamentally, there are two ways to stabilize the expanding foam. It can be physically stabilized by surfactants or chemically by increasing the reaction rate.

Silicone surfactants are particularly effective in foam stabilization, as they form stabilizing polysiloxane monolayers on both sides of the membrane, thereby reducing surface tension. Membrane thinning and bubble ripening are time-dependent processes, and a rapid buildup in viscosity can stabilize the cells. A faster buildup in viscosity can be realized by increasing the reaction rate, for instance, by increasing the catalyst concentration. The effect of catalyst concentration on cell size was observed in rigid foam, where the final foam cell size decreased with increasing catalyst amount [Chapter 7.4.1].

As the density decreases, the cells deform from round to polyhedral shapes. Flexible and rigid foam structures with foam densities below 50 kg/m^3 can be considered polyhedral [5]. However, the foam morphology of rigid and flexible foams is fundamentally different: Rigid foams are closed-cell, whereas flexible foams are open-cell.

Continued polymerization increases the polymer modulus. Eventually, the polymer becomes strong enough to stabilize the foam structure. The foam expansion stops, and the liquid Plateau borders become solid load-bearing struts. In rigid foam, the membranes become "windows".

6.3.5 Surfactants and foam formation

Silicone surfactants are required to prepare low-density PU foam. They are specifically designed to lower surface tension by adsorbing at the liquid-liquid and gas-liquid interfaces, thereby serving two functions: improving the mixing of the reaction components and nucleation and stabilization of the bubbles during foam growth.

Standard non-ionic surfactants such as alkylene oxide block copolymers and substituted nonyl phenols can be utilized to compatibilize the reaction components during mixing. These additives, however, exhibit a limited effect on bubble stabilization during foam growth. Polydimethylsiloxane polyether graft copolymer surfactants show both properties.

Silicone surfactants have been designed for the various PU foam applications, and there is a vast choice of surfactants aiming at specific foam stabilizing effects. The surfactants differ in polydimethylsiloxane backbone length, the number of polyether grafts and their molar masses, and the ethylene oxide-to-propylene oxide ratio [Chapter 2.7.2]. The polydimethylsiloxane chain provides low surface energy, whereas grafted polyether chains ensure compatibility with the reactive liquid. The ratio of polydimethylsiloxane to polyether determines the surface activity of the surfactant; the strength of the surfactant increases with its silicone content. Silicone surfactants effectively reduce the surface tension; at one percent addition, the polyol surface tension is typically reduced by about a third.

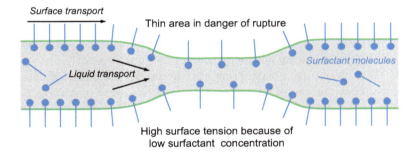

Fig. 6.8: Gibbs-Marangoni mechanism of cell stabilization.

Foam stabilization occurs according to the Gibbs-Marangoni effect. The mechanism can be understood as follows. Let us consider a sudden extension of a cell membrane, whereby the surface area increases and the film thickness locally decreases [Fig. 6.8]. This can occur due to a physical shock or other disturbances in the foam during processing and may lead to membrane rupture and cell collapse. The sudden extension of the film results in a local reduction of the surfactant coverage, and a higher surface tension develops. The generated surface tension gradient initiates a transport process that aims to restore surface coverage, and surfactant molecules migrate from regions of high concentration to those of low concentration. The surfactant-mediated liquid transport mechanism repairs the film perturbation: the surfactant molecules drag back underlying layers of liquid with them into the thin part of the film, thereby restoring the film thickness.

Another aspect of the surfactant monolayer is that it provides viscoelastic surface properties, resulting from the intramolecular interactions of surfactant molecules that counteract the mobility of adjacent molecules. The viscoelastic effect stabilizes the membranes. Increasing the molecular mass of the surfactant at a constant siloxane-to-polyether ratio enhances film elasticity, which results in slower drainage and more stable cell growth.

6.4 Fine-cell rigid foams

Several recent foam studies have focused on rigid foams, particularly understanding their formation [1, 6] and producing fine-cell foams [7, 8].

When preparing hand-mixed water or pentane/water-blown PU foams, mixing speeds of more than 1,000 rpm were required to aerate the reaction mixture properly and obtain fine-cell foams. It was confirmed that spontaneous bubble formation did not occur and that the liberated gases diffused from the reactive liquid toward the initially formed bubbles. The number of bubbles in the liquid was the highest immediately after mixing, amounting to approximately 10^6 cm^{-3}. The initial bubble loss rate was high, and the observed strong initial bubble growth rate resulted from both bubble loss and the generation of cell gases.

Bubble stability improved significantly after the formation of the bubble network. The significant improvement in bubble stability was attributed to the steady increase in cell size and the degree of polymerization. Larger cells result in a lower cell gas pressure, whereas increased polymer molar mass reduces the gas permeation rate and increases membrane stability. The increase in cell size at this stage was largely the result of gas diffusion into the bubbles and an increase in temperature [1, 6].

In principle, when preparing a foam, every nucleus can generate an individual cell [7]. This would imply that increased nucleation leads to finer-cell foams. Fig. 6.9 shows images of a series of standard low-density rigid foams prepared by hand mixing, where the stirring speed was increased stepwise. The nucleation was poor at slow stirring, and the obtained foam exhibits large cells with diameters of approximately 600 μm [Fig. 6.9A]. With increased stirring speed, more nuclei were generated, and the cell size of the foams decreased to approximately 300 μm [Fig. 6.9D]. In this series of foams, the number of air bubbles generated per unit volume in the reactive liquid was of the same order as the cell population density in the final foam. Further increased nucleation, however, did not result in finer cells; the mechanism of which will be discussed in Chapter 7.4.1.

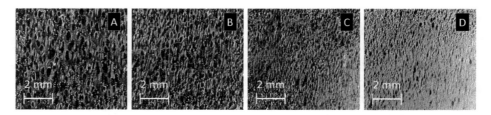

Fig. 6.9: Increased nucleation decreases the cell size from 600 μm to 300 μm (image A to D) (with the kind permission of BASF SE).

It has been known for several decades that adding a few percent of perfluoroalkane to a rigid foam formulation reduces the cell size of the foam. These low-boiling additives are insoluble in polyol and isocyanate, and it was speculated that the insolubility of the additive in the reactive liquid was one of the main reasons for cell size reduction. The mode of action, however, was not understood till recently. The primary pore size-reducing effect of perfluoroalkanes appeared to be the facilitation of air entrainment, resulting in a high initial bubble count. It was observed that adding perfluoro-alkane to a polyol blend containing a silicone surfactant further reduced the surface tension. Complemented by other measurement techniques, it was shown that the per-fluoroalkanes form a macroscopic film on the surfactant layer, altering the nature of the bubble surface and leading to an increased number of initial bubbles [8].

6.5 Foam properties

Foams are classified according to their density into low-density, high-density, and microcellular foams. There are, however, no clear density definitions for the three classes of foams. Low-density flexible foams are open-cell and have densities ranging from 15 to 80 kg/m^3. Microcellular elastomers, such as shoe soles, exhibit densities ranging from 150 to 800 kg/m^3. These foams are largely open-cell below a density of 300 kg/m^3 but closed-cell at higher densities. Rigid foams are always closed-cell. Low-density rigid foams for insulation applications are produced at densities ranging from 30 to 50 kg/m^3. High-density rigid foams, such as those used in artificial wood applications, exhibit densities ranging from 250 to 1,000 kg/m^3.

The cells in foams with a density of approximately 200 kg/m^3 and higher are spherical [Fig. 6.5; page 118]. With decreasing density to approximately 50 kg/m^3, the windows increase in size, whereas the cells remain largely spherical. An ellipsoidal shell with windows or holes provides the best description of the geometry of these higher-density foams. Below a density of 50 kg/m^3, the spherical cells are distorted into polyhedral structures. A typical polyhedral cell consists of 5-sided windows and 14 windows per cell. Deviations occur for rapidly produced fine-cell low-density foams; however, these deviations are not great. The edges, where generally four struts come together, are curved and form a somewhat thicker node than the struts. The struts are the load-bearing elements of the foam, and the thickening at the nodes causes strut bending to be the primary deformation mode during compression [5].

The properties of foams are the combined result of polymer properties and foam structure. The most important foam structural properties are:
- Foam density/foam porosity.
- Cell size and cell size distribution.
- Cell orientation/cell anisotropy.
- Open and close cell content.

Foam density and foam porosity differ in their definition. Foam density is expressed as the mass per unit volume (g/L or kg/m^3), whereas foam porosity (ϕ) is defined as the density of the foam (ρ_f) divided by the density of the solid polymer (ρ_s) and is dimensionless.

$$\Phi = \frac{\rho_f}{\rho_s}$$

The foam density or porosity is the most important structural property. The polymer properties and the amount of polymer per unit volume of foam determine, in a first approximation, the foam's mechanical properties. A power law relationship has been observed for most mechanical foam properties, where X_f represents the foam properties, and X_s the properties of the matrix material.

$$X_f = X_s \cdot \Phi^n$$

The exponent n depends on the property and foam in question and typically lies between 1 and 2 [5].

The polymer mass distribution across the cell windows and struts must be considered for rigid foams. The polymer distribution over struts and windows has been reported to vary between 80:20 and 90:10. Low-density foams exhibit thin cell walls, with thicknesses less than 0.5 µm, and relatively thick struts.

References

[1] C. Brondi, E. Di Maio, L. Bertucelli, V. Parenti, T. Mosciatti, Competing bubble formation mechanisms in rigid polyurethane foaming, Polymer, 228 (2021) 123877.

[2] R. Herrington, R. Turner, W. Lidy, in Flexible polyurethane foams, R. Herrington and K. Hock, Eds. (1997), Chapter 3.

[3] B. Kanner, T.G. Decker, Urethane Foam Formation – Role of the Silicone Surfactant, J. Cell. Plast., 1969, 5/1, 32–39.

[4] R.A. Neff, C.W. Macosko, Simultaneous measurement of viscoelastic changes and cell opening during the processing of flexible polyurethane foam, Rheol. Acta., 35, 656–666 (1996).

[5] N.C. Hilyard, A. Cunningham, Low density cellular plastics, N.C. Hilyard, A. Cunningham, Eds, Chapman and Hall, London (1994) ISBN 0 412 58410 7, Chapter 1.

[6] C. Brondi, M. Santiago-Calvo, E. Di Maio, M.Á. Rodríguez-Perez, Role of Air Bubble Inclusion on Polyurethane Reaction Kinetics, Materials, 2022, 15, 3135.

[7] M. Hamann, S. Andrieux, M. Schütte, B. Telkemeyer, M. Ranft, W. Drenckhan, Directing the pore size of rigid polyurethane foam via controlled air entrainment, Journal of Cellular Plastics, 59(3), 201–214 (2023).

[8] M. Hamann, G. Cotte-Carluer, S. Andrieux, D. Telkemeyer, M. Ranft, M. Schütte, W. Drenckhan, Fluorocarbon-driven pore size reduction in polyurethane foams: an effect of improved bubble entrainment, Colloid Polym. Sci., 302, 585–596 (2024).

7 Rigid foams

The first PU rigid foams, based on TDI and polyester polyols, were introduced into the aircraft industry in the late 1940s. Because of the high NCO content of TDI, the exotherm was high and difficult to control, which limited the industrial application of the early rigid foams. This changed with the introduction of polymeric MDI (PMDI), which has a significantly lower NCO content, the availability of low-cost polyether polyols, and tri-chlorofluoromethane (R-11) as a blowing agent. These starting materials enabled the manufacture of cost-effective, low-density, closed-cell, rigid polyurethane foam with excellent mechanical and insulating properties. Nowadays, rigid foams account for approximately 40% of total PU production and are utilized in various applications, including appliances (such as refrigerators and freezers), insulation boards, metal-faced panels, pipe insulation, and spray foam. The foam density for insulation materials ranges from 30 to 60 kg/m^3, depending on the specific application. The density can be adjusted by varying the amount of the blowing agent.

In most applications, rigid foam is sandwiched between substrates. For instance, a sandwich panel is a composite encompassing a rigid foam slab with metal sheets on both sides of the foam. The key selling feature of PU rigid foams is their excellent heat insulation property. The closed-cell nature of the foam is imperative to keep the insulating gas inside. At the same time, the foam composites show high mechanical load capacity at low densities thanks to the high compression strength of the foam and its strong adhesion to the substrates.

7.1 Rigid foam formulations

The matrix polymer of rigid PU foam is highly crosslinked. The high crosslinking is fundamental to the hardness and dimensional stability of the foam. In conventional rigid PU foam, this can be achieved using high-functional polyisocyanates in combination with short-chain high-functional polyols. The crosslinking can be further increased by using an excess of isocyanate, which is converted to isocyanurate (PIR rigid foam). PIR foams show improved fire resistance.

7.1.1 PU rigid foam

Rigid PU foam is manufactured by reacting PMDI with branched short-chain polyether polyols and water as a chemical blowing agent to expand the foam. Most foam recipes also contain a physical blowing agent, such as pentane.

https://doi.org/10.1515/9783110744583-007

Tab. 7.1: Typical formulation for PU rigid foam.

Component	Amount (p.b.w.)
Polyether polyol	100
Catalyst	1–3
Surfactant	1–2
Flame retardant	0–50
Water (chemical blowing)	1–3.5
Pentane (physical blowing)	0–20
PMDI (index 110–130)	100–200

Tab. 7.1 presents a typical rigid foam formulation. The composition is given in parts by weight, with the amount of polyol set to 100 p.b.w. The polyether polyols are based on propylene oxide and possess secondary hydroxyl groups. The individual polyether polyols have hydroxyl values ranging from 300 to 600 mg KOH/g and functionalities from 2 to 8. Most formulations contain more than one polyol; the polyol blends typically have a hydroxyl value of approximately 450 mg KOH/g and an average functionality of 3 – 5.

The A-component usually consists of several polyols to balance the processing and mechanical properties of the final foam. Processing requires an appropriate viscosity and compatibility with the other reaction components, such as pentane and PMDI. The average functionality and hydroxyl content of the polyol mixture determine the flow of the expanding foam, the exotherm, and the glass transition temperature of the foam.

PMDI is the isocyanate of choice with functionalities between 2.7 and 2.9. In most applications, standard PMDI with functionality 2.7 is used. Rigid PU foams are generally manufactured at an NCO index between 110 and 130. The excess isocyanate may react with urethane to allophanate or trimerize, thus forming additional crosslinks.

Water is the primary chemical blowing agent, and its reaction with isocyanate yields carbon dioxide and a urea linkage. It is possible to produce rigid foams using water as the sole blowing agent; however, this approach has several disadvantages. Water contents as high as 3.5 p.b.w. may be required to achieve foam densities of approximately 35 kg/m^3 when using water as the sole blowing agent. The urea reaction is strongly exothermic, and the high water content causes high reaction exotherms. The high reaction exotherm makes the reaction difficult to control, and, depending on the system and foam size, may even lead to foam scorch. Furthermore, the mechanical properties of the foams are generally poor because of increased urea-to-urethane ratios, which can lead to foam brittleness. The heat-insulating value of all-water-blown foams is fairly poor, and the foams may suffer shrinkage and exhibit poor dimensional stability because carbon dioxide diffuses out more rapidly than air diffuses in. These problems can be overcome by combining water and a physical blowing agent. The water content is then typically reduced to approximately 1 – 2 p.b.w. The reaction exotherm is reduced because of the generation of less reaction enthalpy, which is further reduced by the heat required to evaporate the physical blowing agent.

Nowadays, pentanes (n-, iso-, and cyclopentane) are the most common blowing agents used in the rigid foam industry. They have substituted most other blowing agents because of their low cost, low thermal gas conductivity, low gas diffusion coefficients through the PU matrix material, and low ozone depletion and greenhouse warming potential.

The blowing efficiency of a gas decreases with the increasing molar mass of the blowing agent. Therefore, substituting water for pentane requires significant amounts of pentane to maintain the same density. When using 1.0 – 2.0 p.b.w. of water, up to 20 p.b.w. of pentane is required to achieve a density of 35 kg/m^3. The heat-insulation value decreases with the increasing molecular mass of the blowing agent. Pentane is, therefore, a much better insulator than carbon dioxide and air. Hence, co-blowing with pentane improves the heat-insulation value of the foam. The use of pentane, however, increases the flammability of the foam, which has consequences for its processing and fire properties.

Pentane is highly flammable, and producing pentane-blown systems requires explosion-proof processing equipment. Processors not prepared to invest in this equipment will use the costlier, non-flammable halogenated blowing agents instead. In applications for the building industry, such as insulation panels, incorporating flame retardants is inevitable to meet the required fire rating classifications. Because the flammability classifications are challenging to achieve with standard fire retardants, PU foams have been widely replaced in these applications by pentane-blown polyisocyanurate (PIR) foams. PIR foams exhibit intrinsically improved flame resistance because of isocyanurate linkages.

Tertiary amines are the standard catalysts for rigid PU foams. The most common and work-horse catalyst is DMCHA. The blowing-to-gelling reaction balance can be adjusted by adding gelling or blowing catalysts, such as TEDA and BDMAEE [Chapter 3.2.7].

Surfactants are essential for cell stabilization during foam growth and are pivotal in producing closed and fine-celled foam. A wide range of different polysiloxane-polyether block copolymers is available. The most suitable candidate for a given formulation is generally identified through experimentation.

7.1.2 PIR rigid foam

When more isocyanate is used than is required to convert the reactive groups of the polyol component, then the excess isocyanate can be trimerized to isocyanurate. PIR foams are produced at an index between 200 and 400. These foams are highly cross-linked and possess high glass transition temperatures.

$$3 \quad \underset{R}{\overset{R}{\diagdown}}N{=}C{=}O \longrightarrow$$

With a decomposition temperature of approximately 260 °C, isocyanurate is thermally more stable than urethane, which starts to decompose at around 200 °C [Tab. 3.2; page 50]. Consequently, PIR foams exhibit improved fire resistance compared to rigid PU foams. However, PIR foams show some disadvantages. These include foam brittleness, poor foam adhesion to substrates, and the phenomenon known as the "second rise" during foam expansion. The trimerization reaction starts at temperatures above 60 °C and then proceeds rapidly. The sudden acceleration in isocyanate conversion and the resulting fast release of reaction heat are believed to cause the second rise. A typical formulation for PIR rigid foam is presented in Tab. 7.2.

Tab. 7.2: Typical formulation for PIR rigid foam.

Component	Amount (p.b.w.)
Polyester polyol	100
Catalyst (blowing/gel)	1–3
Catalyst (trimerization)	2–6
Surfactant	1–3
Flame retardant	0–20
Water (chemical blowing)	1–2
Pentane (physical blowing)	15–30
PMDI (index 200–400)	150–400

The isocyanurate structures are temperature-stable and stiff, encompassing the planar aromatic isocyanurate ring and the three aromatic carbon skeletons of MDI. At higher indices, interconnected isocyanurate/aromatic structures can form, providing additional stability and rigidity to the polymer. With increasing index, the glass transition temperature of the polymer matrix material increases, which can lead to foam brittleness. Using low-functional polyols with increased molar mass can reduce the brittleness. PIR foams are generally produced using aromatic polyester polyols with functionalities of approximately two and hydroxyl values of 200 – 300 mg KOH/g. These polyols reduce foam brittleness, whereas their aromatic content further improves fire resistance.

The amount of water in PIR formulations should be kept low. Water inhibits the trimerization reaction, limiting its use in PIR foams. Therefore, to achieve the same foam densities, the amount of physical blowing agent is usually somewhat higher than that in rigid PU foam formulations.

The catalysis of PIR rigid foams is more complex than that of rigid PU foams as three competing reactions must be balanced: blowing, gelling, and trimerization. Alkali metal or quaternary ammonium salts of carboxylic acids are efficient trimerization catalysts. The most common catalysts are potassium acetate and potassium 2-ethyl hexanoate.

The amine 1,3,5-tris[3-(dimethylamino)propyl]hexahydro-1,3,5-triazine is used as a co-catalyst in the trimerization reaction. Amine catalysts such as DMCHA are added to catalyze the urea and urethane reactions. Other amine catalysts may be added to optimize the rise profile further.

The fire resistance requirements in construction applications can generally be met with PIR foams using relatively low amounts of phosphorus fire retardants, allowing the production of halogen-free foam.

7.2 Glass transition temperature

Rigid PU foams are produced from highly branched polyols with equivalent masses between 100 and 200 g/mol and PMDI. Every branched molecule in the formulation forms a crosslink in the polymer, resulting in an average molar mass between crosslinks of 200 – 400 g/mol. The high crosslink density and low equivalent mass of the polyol branches prevent phase separation, resulting in the formation of an amorphous polymer glass. The polymer is hard and brittle at ambient temperature and softens at the glass-rubber transition.

The T_g can be determined experimentally from the modulus-temperature curve of the polymer recorded by DMA. Fig. 7.1 illustrates a typical plot of the storage modulus (G') versus temperature for a rigid PU foam. Extrapolating the modulus-temperature lines from the rubbery and glassy state gives the T_g at the intersection. The glass-rubber transition is broad; however, the T_g can be determined relatively accurately, with an error of about 5 °C. To achieve sufficient high-temperature stability of the foam, the T_g of PU rigid foam polymers must be higher than 140 °C.

The T_g increases with increasing crosslinking density. The polymer crosslink density increases with increasing functionality of isocyanate and polyol and decreasing equivalent mass of the polyol. The T_g is strongly dependent on the choice of polyol, whereas the functionality of the PMDI has only a minor effect on it. Restricting the mobility of the flexible polyether segment appears to have a bigger impact on the T_g than increasing the functionality of the stiff MDI segment.

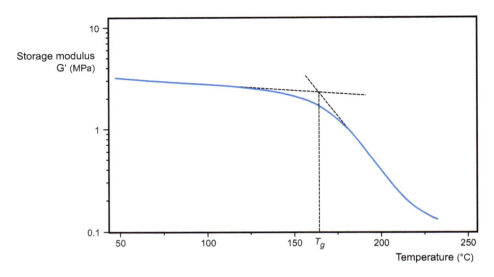

Fig. 7.1: Temperature dependence of the storage modulus of typical rigid PU foam.

The effect of functionality and equivalent mass of the polyol on the T_g is presented in Tab. 7.3. The foams were prepared using polyol and standard PMDI at an index of 100. The first series is based on oxypropylated glycerine polyols with equivalent masses from 90 to 160 g/mol. As the equivalent mass of the polyol increased, the amount of PMDI had to be reduced to achieve an index of 100. The mass fraction of "stiff" aromatic building blocks decreases, resulting in a decrease in T_g.

Tab. 7.3: Influence of equivalent mass (hydroxyl value) and functionality of the polyol on the glass transition temperature.

Series 1 (f_n = 3)	EM (g/mol)	M (g/mol)	OHv (mg KOH/g)	T_g (°C)
	90	270	625	130
	100	300	560	120
	120	360	465	100
	160	480	350	70
Series 2 (EM = 100 g/mol)	EM (g/mol)	M (g/mol)	OHv (mg KOH/g)	T_g (°C)
	2	200	560	75
	3	300	560	120
	4	400	560	145
	6	600	560	175

In the second series, the functionality (f_n) of the polyol was varied from 2 to 6 at a constant equivalent mass of 100 g/mol. With increasing functionality, the mobility of the polyol segment gets increasingly restricted, and the T_g increases. So, an increase in polyol functionality and a decrease in its equivalent mass increase the polymer's T_g.

As the index increases, the mass fraction of isocyanate increases, and additional crosslinks, such as allophanate and isocyanurate, are formed. The increased crosslink density and content of stiff aromatic chain segments increase the T_g. Above an index of 200, the T_g can exceed 200 °C. The experimental determination of the T_g of these foams is challenging because degradation reactions may begin to influence the measurement.

7.3 Foam formation

Rigid foams are produced by reacting two components: the isocyanate component and the polyol component, containing the polyol and additives. In the lab, simple hand-mixing is used to prepare a cup foam, whereas in production, the foams are machine-produced using low or high-pressure mixing machines.

The foaming reaction involves complex changes. Density changes accompany the polymer's transition from liquid to rubbery and, finally, to rigid. All these occur under conditions of rapid temperature changes. The foam rise profile can be captured by recording the height of the expanding foam over time. However, the most commonly used parameters to characterize the expansion characteristics are the experimentally determined times to creaming, gelling, and end of rise, directly measured on the expanding foam.

7.3.1 Reaction profile and cure

After mixing, several characteristic times are recorded to describe the reaction profile of the expanding foam. The reaction profile of a free-rise hand-mixed cup foam using a typical PU formulation is discussed.

The time starts when the two components are being mixed. The first characteristic time is the "mixing time", for a cup foam typically 5 – 10 s. After mixing, the foam starts to expand, and the liquid becomes creamy. The time when this happens is referred to as cream time. The foam continues to expand, and the polymer molar mass increases. At a conversion of about 50%, the reaction mixture forms a chemically crosslinked gel. Strings can be withdrawn from the foam by dipping and pulling a pin in and out of the foam. The time at which the first string can be pulled is referred to as the string or gel time. With progressing crosslinking, the foam strength increases. When the polymer strength is high enough to withstand the difference between the cells' external (atmospheric) and internal pressures, the foam rise stops, marking the end of rise time.

The cream, string, and end of rise time characterize the reaction profile. For the present PU foam, these were 15, 60, and 90 s, respectively. The foam density was measured to be 35 kg/m^3.

Fig. 7.2 illustrates the height and the rise rate (the first derivative of height to time) as a function of time. The height profile, generally referred to as the rise profile, exhibits an S-curve, and the reaction times are indicated.

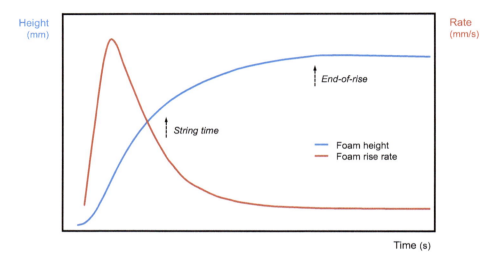

Fig. 7.2: The foam height (blue) and the rise rate (red) versus time.

The above-described cup foam test can be extended by recording the temperature and measuring hardness over time. These additional measurements indicate that after the end of rise, the reaction continues – the temperature and foam hardness steadily increase. The maximum temperature (T_{max}) of 170 °C was reached after 4 min, and the foam showed a hardness plateau after 8 min. The final foam hardness and density are reached after a few days. DMA showed that the T_g (in terms of the TTT-cure diagram, this corresponds to $T_{g\infty}$) of the core material was 140 °C.

The observation that achieving a hardness plateau takes longer than reaching the maximum temperature can be explained qualitatively using the "Time-Temperature-Transformation diagram" (TTT diagram) [Chapter 3.7.2.].

The TTT diagram illustrates the major physicochemical events in polymer formation, highlighting gelation, vitrification, and full cure.

Fig. 7.3 illustrates the three curves for gelation, full cure, and vitrification, as well as the temperature profiles of the present foam measured in the core (T_{core}) and 1 cm below the foam surface ($T_{surface}$).

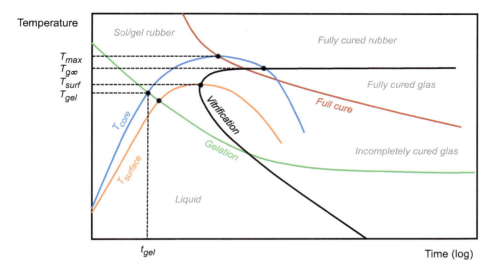

Fig. 7.3: Schematic TTT cure diagram of polyurethane adopted from [1], with temperature development of the foam core (T_{core}, blue) and 1 cm below the foam surface ($T_{surface}$, orange).

The core foam reaches gelation (temperature crosses the green curve at t_{gel}) at approximately 50% conversion. In the gelled state, the reactivity remains high, and the polymer reaches full conversion (temperature crosses the red curve at T_{max}). The polymer is then in its rubbery state because $T_{max} > T_{g\infty}$. With time, the reaction heat dissipates, and the temperature decreases. Given sufficient time, the temperature will decrease below $T_{g\infty}$ (temperature crosses the black curve). At this point, the vitrification of the foam begins, resulting in an increase in foam hardness.

The temperature measured at the surface of the foam nearly matches the core temperature until gelation. Thereafter, the reaction rate slows down, and the surface temperature begins to fall behind that of the core. The surface material reaches a T_g equal to the reaction temperature ($T_{surface}$ crosses the black line), which marks the onset of vitrification. Continued vitrification freezes the reaction, which may prohibit full cure ($T_{surface}$ does not cross the red curve because the reaction ceases). Due to poor conversion, the surface layer may be brittle and exhibit poor strength.

PIR rigid foam rise profiles differ from those of PU foam. In the PIR reaction, water and hydroxyl groups are first converted to urea and urethane. Subsequently, the excess isocyanate is converted into allophanate and isocyanurate, but this predominantly occurs during the later stages of the reaction when the temperature has exceeded 60 °C. The isocyanurate formation proceeds rapidly and boosts the reaction exotherm. As a result, the foam expansion suddenly accelerates, a phenomenon known as the "second rise". The second rise is visible as a second maximum in the rise rate curve [Fig. 7.4].

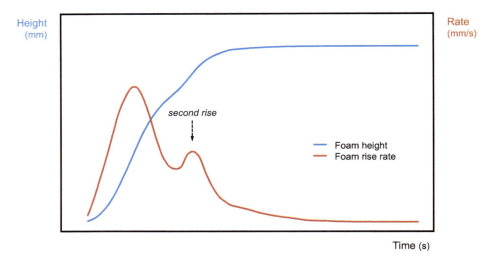

Fig. 7.4: Rise profile of PIR rigid foam.

Because of the high crosslinking and aromatic content, the $T_{g\infty}$ of a PIR foam is high and may exceed 200 °C. The exotherm of a PIR foam is generally somewhat higher than a PU foam; however, in general, $T_{max} < T_{g\infty}$. This can lead to an incomplete cure, which is likely to be evident at the surface, but may also occur in the core of the foam.

Isocyanurate formation generates crosslinks and stiffens the polymer. Shortly after the trimerization reaction kicks in, the polymer starts to vitrify, and simultaneously, the expansion rate increases significantly (the second rise). When producing PIR foam boards or panels on a laminator, the sudden onset of vitrification, combined with the high foam expansion rate, can lead to the formation of shear layers or even cracks in the foam. The shear layer is observed as a layer of tilted cells in the middle of the foam. The shear layer is a foam defect; the layer of kicked cells reduces the compression strength values of the foam.

Selecting the optimal combination of starting materials, refining the catalyst package to mitigate the second rise, and optimizing the processing conditions may help overcome the challenges associated with PIR foam processing.

7.4 Properties

PU rigid foams exhibit a unique combination of low thermal conductivity, excellent mechanical properties, and good surface adhesion to substrates. Changing the formulation enables the manufacture of a wide range of PU rigid foams with varying foam densities and properties. Depending on the application, the densities range from 30 to 50 kg/m³ for heat insulation applications, up to 300 kg/m³ for modeling blocks, and 800 kg/m³ for artificial wood applications. Most PU rigid foams are closed-cell; however, open-cell foams can be produced and used in applications such as vacuum insulation panels.

7.4.1 Cell structure

The cell structure of rigid foams is characterized by three parameters: the closed-cell content, the cell size, and the cell anisotropy. Fig. 7.5 illustrates the typical closed-cell foam structure of low-density rigid PU foam. The cell size of a rigid foam typically ranges from 150 to 300 μm. The struts configure the cell shape, and the cell windows confine the cell. In low-density foams for heat insulation, the windows exhibit thicknesses of less than 0.5 μm; however, the barrier properties of the thin windows are sufficient to retain the gases of the physical blowing agent within the foam. The struts are the primary load-bearing elements, comprising approximately 80 – 90% of the polymer material.

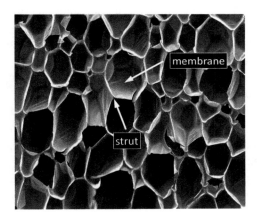

Fig. 7.5: PU rigid foam structure. Note the cell orientation in rise direction of the foam (from the bottom to the top) (with the kind permission of Covestro AG).

Closed-cell content

When the fraction of open cells in rigid foam increases, the foam's heat insulation value decreases, because open cells allow air ingress and air has a higher thermal conductivity than the used blowing agents. Therefore, a high closed-cell content is pertinent for good heat-insulating properties.

During foam expansion, the cell size increases, and the membranes become thinner and less stable, which potentially may lead to the formation of open cells. Adding a physical blowing agent at reduced water content aids cell stability. The reduced average hydroxyl value of the polyol component reduces the exotherm, and the evaporation of physical blowing agents requires heat. The reduced temperature lowers the cell gas pressure, improving cell stability near the end of rise. In contrast, adding more water and less physical blowing agent increases the temperature and cell pressure, thus destabilizing the cells. The formation of insoluble urea particles [Chapter 6.3.4] can further compromise membrane stability. Therefore, foams blown with high amounts of water are more prone to form open cells.

The closed-cell content of rigid foams is determined using ISO 4590, based on gas displacement. This measurement uses a gas pycnometer, and calculations are based on Boyle's law. The closed-cell content of rigid PU foams generally ranges between 85 and 95%.

Cell size

The cell size of rigid foam depends on the number of nuclei formed during mixing and the loss of bubbles during foam expansion. The bubble decay rate is especially high shortly after mixing.

During mixing, small gas bubbles are generated in the reaction mixture. These gas bubbles serve as nuclei for the formation of foam cells. In principle, every nucleus can form a cell; therefore, increased nucleation should result in smaller cells [Fig. 6.9; page 122]. However, the cell size reduction by gas nucleation has a limit: above a critical number of nuclei, the cell size is not further reduced. Increased nucleation inherently leads to the formation of smaller gas bubbles with higher internal gas pressures and smaller inter-bubble distances. As a result, the rate of bubble loss increases, which rapidly reduces the number of initial small bubbles. Consequently, a further increase in nuclei above the critical number does not lead to smaller cell sizes.

Bubble degradation is a time-dependent process and is dependent on the liquid viscosity. With increasing reactivity, the viscosity build-up is accelerated, which, in turn, stabilizes the bubbles and reduces the average cell size of the foam. There are, however, practical limitations to the extent to which the reaction rate can be increased. Consequently, the reduction in cell size by increased nucleation and reaction rates is limited. Currently, the smallest cell sizes achieved using conventional nucleation technology range from 150 to 200 µm.

Cell anisotropy

The foam cells of rigid foams are anisotropic and oriented in the direction of foam rise, as shown in Fig. 7.5 and depicted in Fig. 7.6. The cell orientation is determined by the physical restriction of the foam during expansion and the gelation characteristics of the PU polymer.

Consider preparing free-rise rigid foam in a vertical cylinder. The molecular mobility before gelation is high, and the influence of the cylinder wall on the foam flow is small. After gelation, the foam attains elastic properties, and further expansion occurs largely through polymer stretching. The cylinder wall impedes the foam flow with further expansion, and the cells begin to stretch. The cell orientation increases with a decreasing cylinder diameter and an increasing functionality of the polyol and isocyanate.

The anisotropy in cell shape R is measured by the ratio of the mean length in the rise direction (r_1) to that in the perpendicular plane (r_2).

$$R = \frac{r_1}{r_2}$$

R for low-density rigid PU foams was reported to vary between 1.1 and 1.5, and tended to increase with cell size and decrease with density. The ratio of the compressive strength (σ) perpendicular and parallel to the rise in this series of foams was measured to be between 1.4 and 2.0 [2].

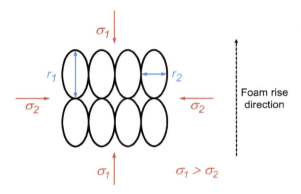

Fig. 7.6: Cell anisotropy in rigid foams ($\sigma = F/A$).

7.4.2 Thermal conductivity

Considerable effort has been made to understand heat transfer mechanisms, and heat conductivity models have been proposed to predict heat conductivity from first principles [3–5]. The heat conductivity measurement is based on the heat flow principle and is performed according to ISO 8301 (heat flow meter) or ISO 8302 (guarded hot plate).

When a material is placed between two environments of different temperatures, heat flows through the sample from the hot (T_2) to the cold (T_1) environment [Fig. 7.7]. The thermal conductivity is calculated from the measured heat flux q, the applied temperature difference (T_2-T_1), and the thickness of the foam (L). The heat flux q is the energy flow per unit time per unit area (W/m^2). The heat conductivity of the solid is then calculated by multiplying the heat flux by the sample thickness and dividing by the temperature difference.

$$\lambda = q \cdot \frac{L}{T_2 - T_1}$$

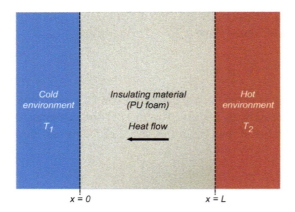

Fig. 7.7: Heat transfer through a plane solid material.

The constant Lambda (λ) represents the material's thermal conductivity, which is called Lambda value or k-factor. The λ-value is a material constant expressed in W/(m·K). It is often multiplied by 1,000 to obtain two-digit heat insulation values. The unit is then mW/(m·K), or "milliwatt per meter Kelvin".

The λ-values of common insulating materials used in the construction industry are summarized in Fig. 7.8, showing that PU is the best commercial insulating option with λ-values between 20 and 25 mW/(m·K).

Heat transfer occurs through conduction within the PU matrix and cell gas, and by radiation. Because of the small cell sizes, increased gas conduction by cell-interior convection does not occur. Convection will only begin when the cell size reaches approximately 5 mm [4].

The heat conductivity of a foam can be expressed by a superposition of each of the mechanisms taken separately.

$$\lambda - value = \lambda_{solid} + \lambda_{radiation} + \lambda_{gas}$$

In a typical low-density PU foam filled with a low-conductivity cell gas, the heat transfer through the gas accounts for more than half of the total heat transfer. The solid conduction and radiation contributions are about equal.

The foam density determines the heat transfer through the PU matrix (λ_{solid}). The λ-value of compact PU is approximately 200 mW/(m·K). By proportion, the λ_{solid} in rigid foams with 35 kg/m³ density would amount to 7 mW/(m·K); however, in a foam, λ_{solid} is only about half of that. Foams transfer heat less efficiently by conduction because only part of the solid phase is oriented toward the heat flux. The λ_{solid} increases with density by approximately 1 mW/(m·K) per 10 kg/m³. A foam with a density of 35 kg/m³ typically exhibits a λ_{solid} of about 3 mW/(m·K).

The radiation term ($\lambda_{radiation}$) depends on the cell size and foam density. It describes the heat transfer by electromagnetic waves generated by the thermal motions

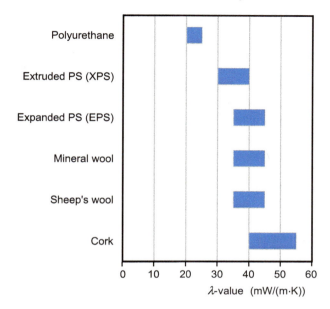

Fig. 7.8: Comparison of thermal insulating properties (λ-value) of different materials used in the building industry.

in the material. This term can be described using the Stefan-Boltzmann law and depends on the third power of the absolute temperature [4]. In foams, the radiation emitted from a cell strut can be reflected and absorbed by a nearby strut, which, in turn, emits radiation, and this process repeats. At a constant density, $\lambda_{radiation}$ increases with the cell size because the number of reflection barriers per unit volume decreases. The increase in $\lambda_{radiation}$ with cell size for a foam with a density of 35 kg/m³ was calculated to be approximately 7.5 mW/(m·K) per mm cell size [3]. This value aligns well with the experimentally determined 6.6 mW/(m·K) per mm cell size [6].

At a constant cell size of 300 µm, $\lambda_{radiation}$ remained largely constant in the density range from 100 to 60 kg/m³ with a value of approximately 2.5 mW/(m·K). However, it increased with further decreasing density to 35 and 25 kg/m³ from 3.5 to 4.0 mW/(m·K), respectively. At low densities, $\lambda_{radiation}$ increases because the fine and thin struts are less effective in reflecting the radiation. The increase in $\lambda_{radiation}$ at low densities results in a minimum in λ-value at foam densities of about 28 kg/m³ [3].

The sum of λ_{solid} and $\lambda_{radiation}$ for a standard rigid PU foam with a density of 35 kg/m³ and a cell size of 300 µm amounts to 6.5 mW/(m·K). The λ-value of a foam filled with gas is this number plus the gas contribution.

The gas conductivity is a function of the mean free path length, which decreases with increasing molecular diameter [5]. As a result, the thermal conductivity decreases with increasing molecular mass. The λ_{gas} increases [Tab. 7.4] from 10.8 mW/(m·K) for HFC-365mfc to 24.7 mW/(m·K) for air. A standard air-filled rigid foam is then predicted to have a λ-value of 31.2 mW/(m·K), whereas a foam filled with HFC-365mfc would have a λ-value of 17.3 mW/(m·K) [7].

Generally, a mixture of blowing gases, such as the often-used combination of carbon dioxide and pentane, is used to expand the foam. The thermal conductivity of gas mixtures with significantly different molecular masses is poorly represented by the average of the separate gases, weighted by their respective mole fractions. The mean free path length of a gas component is altered when a second component is present. Carbon dioxide in the presence of pentane will have a reduced contribution to the conductivity because the larger molecules will retard the motion of the smaller molecules. The mixture will, therefore, have a lower λ-value than the average weighted by their mole fractions. The λ-value for mixtures of gases can be reasonably predicted, for instance, using the Lindsay-Bromley relationship [4].

Tab. 7.4: Gas thermal conductivity of different cell gases at 10 °C.

Cell gas	Molecular mass (g/mol)	λ-value (mW/(m·K))
Nitrogen	28	24.6
Oxygen	32	24.9
Carbon dioxide	44	15.3
n-Pentane	72	13.7
HFC-134a	102	12.3
HFC-245fa	134	11.3
HFC-365mfc	148	10.8

Vacuum insulation panels

The contribution of gas conductivity is the largest of the three factors determining the λ-value of standard PU foam. Significantly improved λ-values would be achieved without any gas in the cells, which is practically realized in vacuum insulation panels (VIP). They are produced by sealing a slab of open-cell rigid foam with a gas-impermeable aluminum foil and evacuating the panel to pressures below 0.01 bar. The open cell content of foams used for VIPs must be high because the cell gases entrapped in closed cells will diffuse out over time. With possible foam emissions or air leakage through the foil, the pressure increases and concomitantly, the λ-value of the VIP. An absorptive material, or "getter", is incorporated into the panel to counteract pressure increases, thereby warranting a service lifetime for the VIP panel of about 10 years. The λ-value of VIPs is approximately 7 mW/(m·K), essentially the sum of $\lambda_{radiation}$ and λ_{solid}. VIPs are used in manufacturing appliances and low-temperature transportation boxes for medical purposes.

Lambda-value aging

The λ-value of PU foam increases with time because blowing agents diffuse out and air diffuses in. The carbon dioxide diffusion rate through the foam is high, intermediate for air, and low for physical blowing agents.

The differences in partial pressures of the gases in the cells and the environment drive the diffusion processes. This process continues until the partial pressures of the gases in the cells and the environment are equal. Carbon dioxide escapes completely as its partial pressure in air is close to zero. Air migrates into the cell until its partial pressure reaches 1 bar. The physical blowing agent has a very low diffusion coefficient; it diffuses out, but the process may take decades [4].

Let us consider the aging characteristics of a PU foam produced from a recipe containing 1.5 p.b.w. of water and 14 p.b.w. of n-pentane on 100 p.b.w. of polyol. This yields a foam with a density of 35 kg/m^3 and an initial cell gas composition of 30:70 molar ratio of carbon dioxide and n-pentane. At the end of the rise, the foam temperature and the cell gas pressure were measured to be 160 °C and 1.2 bar. Using the ideal gas law and assuming that the cell volume remains constant upon cooling yields a cell gas pressure at 20 °C of 0.81 bar. Multiplying the molar ratio of the two gases yields their partial pressures: 0.24 bar for carbon dioxide and 0.57 bar for n-pentane.

The diffusion processes in a foam are complex and involve multiple factors. A simplified model consisting of a single isolated PU foam cell surrounded by air (N_2) was used for the calculations to demonstrate the basic principles of foam aging [5]. The advantage of this model is that the diffusion processes only depend on the partial pressure differences between the gases in the cell and the environment. The calculations used the gas diffusion coefficients from reference [5], which were normalized to 1 for air to demonstrate the relative differences [Tab. 7.5].

Tab. 7.5: Gas diffusion coefficients for PU polymer at 20 °C.

Gas	Diffusion coefficient (m^2/s)	Diffusion coefficient (normalized)
Air (N_2)	$4.66 \cdot 10^{-12}$	1
Carbon dioxide (CO_2)	$1.20 \cdot 10^{-10}$	25.75
n-Pentane	$3.17 \cdot 10^{-14}$	0.0068

The time axis in the model is dimensionless, but the calculated pressures are real. The calculation results are given in Fig. 7.9. The effusion of carbon dioxide occurs fast. Because air diffuses in at a slower rate, a minimum in the total cell gas pressure is reached after a relatively short time. As more air diffuses in, the total pressure increases and, over time, reaches a plateau value of approximately 1.5 bar. The partial pressure of n-pentane does not decrease significantly over the given time interval. Therefore, the gas pressure increment is the direct consequence of air ingress.

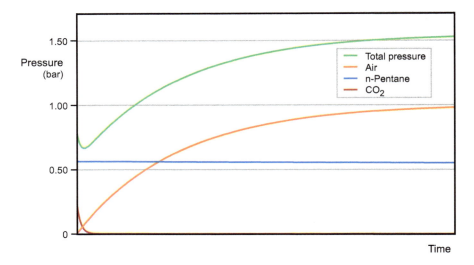

Fig. 7.9: Partial cell gas pressures of carbon dioxide, air (N_2), and n-pentane and the total gas pressure with time of the example pentane-water blown rigid PU foam.

The compression hardness of rigid PU foams is measured on 50 mm cubes cut from a larger block. The compression hardness of (partially) water-blown foams shows a minimum after approximately 10 days. This minimum results from the minimum in gas pressure of the foam.

Fig. 7.10 illustrates the corresponding development of the λ-value with time based on the calculated cell gas compositions shown in Fig. 7.9. The combined contribution of solid conduction and radiation was taken as 6.5 mW/(m·K), whereas the gas contribution was calculated using simple molar proportions. The thermal conductivity curve follows essentially the air ingression curve of Fig. 7.9 and reaches a plateau value. The predicted initial and final λ-value were 20.7 and 27.2 mW/(m·K), incrementing 6.5 mW/(m·K).

Given sufficient time, PU foams not protected by facers typically exhibit an increase in thermal conductivity between 4 and 7 mW/(m·K). In practice, however, foams for insulation applications are sandwiched on both sides by substrates, which will retard the aging process. Impermeable facings can effectively prevent aging during the panel's service life.

7.4.3 Dimensional stability

The foam's dimensional stability over a wide range of temperatures is relevant for many applications, such as in the construction industry, where the operating temperature for PU rigid foams can range from well below 0 °C to above 80 °C.

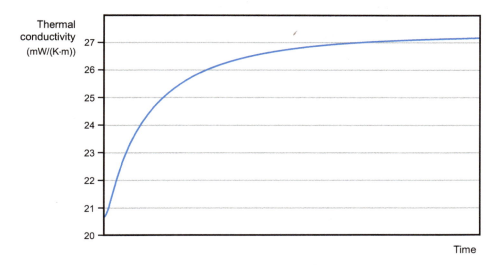

Fig. 7.10: Development of the λ-value with time for the example pentane-water-blown PU rigid foam.

The dimensional stability test measures the tendency of the foam to swell or shrink under service conditions. Foam slabs (100 mm × 100 mm × 25 mm) are subjected to specified temperatures for a given duration, after which the changes in the three orthogonal directions are measured. The samples are typically held at low (-18 °C) and high temperatures (+70 °C or +100 °C) for 24 h (ISO 2976) and the dimensions of the slabs are measured when the foam has reached room temperature. Some shrinkage (≤1%) in each direction is acceptable.

When the pressure difference between the gas pressure in the cells and the atmospheric pressure exceeds the strength of the polymer, it will deform. This can happen at both high and low testing temperatures. At higher temperatures, the gas pressure increases. The foam slab may swell when the polymer has a low crosslink density and a low glass transition temperature. The physical blowing agent may condense at low temperatures, reducing the cell pressure. Although the polymer is rigid at those temperatures, it may shrink during the test.

7.4.4 Mechanical properties

The mechanical properties of PU foams, including flexural, tensile, and compressive strength, primarily depend on the crosslinking of the polymer matrix, the cell structure, the cell gas pressure, and the foam density. Compressive strength is the most important of the three properties and is measured according to ISO 844. The flexural and tensile properties are measured per ISO 1209 and ISO 1922, respectively. The reported strength values of foams are always measured perpendicular to the rise direction of the foam unless otherwise stated.

The compressive strength of rigid foams is determined using foam cubes with dimensions of 50 mm × 50 mm × 50 mm. The foam is positioned between two parallel plates and then compressed at a constant rate. Fig. 7.11 illustrates the relationship between compression stress and compressive strain schematically. Up to approximately 5% compression, the cells undergo homogeneous and reversible deformation, exhibiting a nearly linear strain-stress relationship with a slope of E^*, the Young's modulus of the foam. The cells irreversibly collapse at higher compression, resulting in a plateau value in compressive strength. When a maximum is reached before 10% compression, the highest value in compressive strength is taken; otherwise, the compressive strain at 10% compression is taken as compressive strength. The compression curve may exhibit an upturn at high compression due to densification, which is not relevant to the test results.

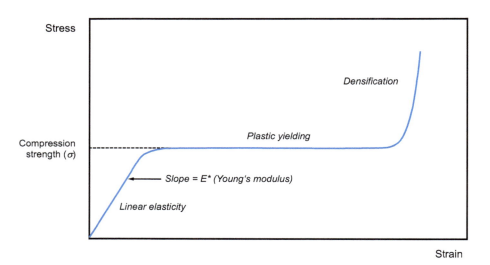

Fig. 7.11: Typical stress-strain curve of PU rigid foam measured under compression.

Fig. 7.12 illustrates the typical flexural, tensile, and compressive strength values of rigid PU foams versus foam density. The strength values appear to increase linearly with foam density over the investigated density range; however, a more detailed analysis reveals the following relationships with the density:

Compressive strength $\sim \rho^{1.5}$
Tensile strength $\sim \rho^{1.35}$
Flexural strength $\sim \rho^{1.75}$

The compression strength correlates to the density to the power 1.5. This value aligns with the theoretical collapse stress of open-cell foams that exhibit plastic collapse under compression, such as rigid PU foams [8]. The close agreement suggests that the contribution of the cell membranes to compression is small.

The tensile strength and density relate to the power 1.35. Rigid foams show brittle failure under extension. The found exponent agrees well with the theoretical value of 1.33 for open-cell brittle foams, suggesting that the cell membranes contribute little to the tensile strength and that the fracture toughness is independent of cell size [8].

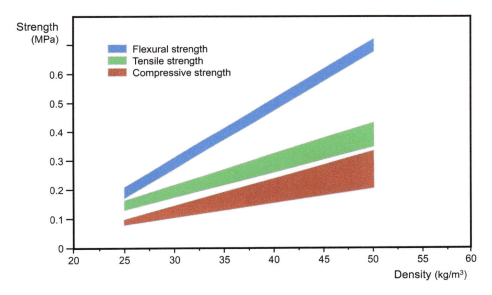

Fig. 7.12: Flexural, tensile, and compressive strength versus foam density (with the kind permission of Covestro).

7.4.5 Flame resistance

Flammability is an important property for rigid PU foams in the construction industry. Most countries have specific fire testing protocols for insulation materials that must be followed in international trade [Tab. 7.6]. Two representative tests, the small burner and the SBI test, are discussed.

Tab. 7.6: Examples of national burning tests.

Country	Standard	Test	Classification
Germany	DIN 4,102–1	Small burner	B2, B3
Germany	DIN 4,102–15/16	Fire shaft	B1
France	NF P 92–501	Epiradiateur	M1 – M3
United Kingdom	BS 476		Class 1
Europe	EN 13,823	Single burning item	A2, B – D
			s1 – s3 / d0 – d2
USA	ASTM E84	Steiner tunnel	A – C

Small burner test

The small burner test (DIN 4,102–1) simulates the reaction to fire of foam coming into contact with a burning match. Depending on the test results, the materials are classified into Class A1 or A2 ("non-combustible materials"), B1 ("not easily flammable"), B2 ("flammable"), or B3 ("easily flammable"). Only inorganic materials can meet the classifications A1 and A2. The minimum classification for all materials brought to the building site in Germany is class B2, which corresponds to the burning behavior of wood.

Fig. 7.13: Test setup according to DIN 4,102–1 (with the kind permission of Prüfinstitut Hoch).

Fig. 7.13 illustrates the test setup for the small burner test. The specimen is placed into a holder and ignited from the side using a pilot flame. The flame is removed after 15 s. The B2 classification is met when the highest flame observed within 20 s after ignition does not exceed 150 mm and no drippings are observed.

Single Burning Item

Most European construction products are tested and classified using the Single Burning Item test (SBI; EN 13,823). The Single Burning Item test method is used to determine the reaction-to-fire behavior of building products when exposed to a thermal attack by a single burning item. Two sides of a corner are covered with the material to be tested, e.g., PU lamination boards. The height is 1.5 m; the widths of the two sides are 1.0 m and 0.5 m. A 30 kW propane gas burner is placed at the bottom of the corner, and the assembly is exposed to the flame for 20 min [Fig. 7.14].

Fig. 7.14: Single Burning Item (SBI) test (with the kind permission of Prüfinstitut Hoch).

During that time, the combustion gases are monitored, and the total heat and smoke production are measured. The results are given as FIGRA ("Fire Growth Rate"), SMOGRA ("Smoke Growth Rate"), and THR$_{600}$ ("Total Heat Release over the first 10 min). In addition, the formation of drippings is monitored. Depending on the results, the material is classified by FIGRA (Class A2, B, C, D), SMOGRA (s1 – s3), and drippings (d0 – d2).

7.5 Applications and processing

Thermal insulation is the key property in almost all rigid PU foam applications. The good adhesion of PU foam to various facing materials enables the production of construction panels for the building industry, including sandwich panels with rigid facings, insulation boards with flexible facings, and spray foam for wall and roofing insulation. In the appliance industry, PU rigid foam is used as a thermal insulation material for domestic and commercial refrigerators, freezers, and boilers. Industrial applications include insulating tanks, containers, and industrial piping. In the transportation industry, PU foams are used for the thermal insulation of refrigerated vehicles on road and rail, including freight containers. The formulations for each application must be fine-tuned to meet the specific requirements regarding processing and physical properties of the foams.

7.5.1 Rigid slabstock

Rigid slabstock foam (bun stock) can be produced by discontinuous and continuous processing. The components are typically mixed using low-pressure technology, which has the advantage of allowing highly viscous components, such as filler-containing polyols, to be processed.

Discontinuous processing
The liquid reaction mixture is poured into a wooden box with dimensions of approximately one cubic meter. The box surfaces are covered with release paper or treated with release agents to facilitate demolding. To reduce dome formation, the mold is covered with a floating lid [Fig. 7.15] that abuts against a stopper at the high end of the box, thereby slightly compacting the foam.

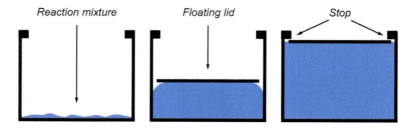

Fig. 7.15: Foaming in a box mold with a floating lid.

Continuous processing
Fig. 7.16 illustrates a conveyor belt line for producing rigid slabstock foam. The line is paper-protected to avoid contact between foam and the machine. The reaction mixture is continuously poured onto the moving conveyor belt. The foaming pressure during the rise is relatively high. To counteract the lateral expansion, the conveyor is equipped with moving side belts. Dome formation can be prevented by applying floating lids, also known as the "flat-topped slab method". The rigid foam slab ex-conveyor is cut into

Fig. 7.16: Conveyor belt for the continuous production of rigid slabstock foam (with the kind permission of puren GmbH).

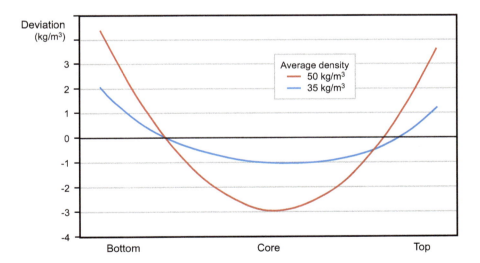

Fig. 7.17: Schematic density distribution in rigid slabstock foams.

blocks of the desired length. The foams are produced at densities ranging from 30 to 200 kg/m³; the block widths typically vary between 1.0 and 1.5 m, and the height ranges from 0.4 to 1.0 m.

The foams exhibit a density distribution from the top to the bottom. The temperature in the middle of the block is higher than that of the periphery because of heat losses. The temperature gradient over the bun creates a density gradient, with the lowest densities at the highest temperatures. The deviation in density increases with the overall density of the foam. In the present example, a density increase of 35 to 50 kg/m^3 led to a deviation increase from approximately 3 to 6 kg/m^3 [Fig. 7.17].

7.5.2 Laminated boards

PU allows the manufacturing of in-situ laminated boards because the adhesion to the facings is automatically obtained during the foaming process when the tacky foaming reaction mixture comes in contact with the substrates.

Discontinuous processing
Discontinuous processing gives the manufacturer more design flexibility and is preferred for small production volumes or specialty applications, such as garage doors. The panels are produced in hydraulic presses [Fig. 7.18] with facings that are secured by applying a vacuum through small orifices in the mold surface. The molds are heated to 30 – 40 °C to promote surface curing and ensure good adhesion to the facings. Single-point injection can be used for small molds, but lance or withdrawal techniques are required for long molds, which may be up to 20 m, for example, when manufacturing panels for refrigerated trucks. In the latter case, a lance equipped with the mixing head is pulled from one end to the other, uniformly filling the mold with the reactive mixture.

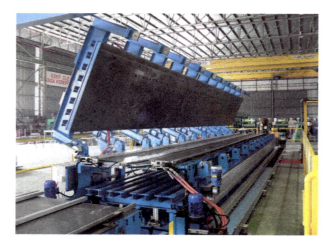

Fig. 7.18: Open pour press for discontinuous manufacturing of laminated boards (with the kind permission of AutoRIM Ltd).

The lance technology requires long pouring times of up to 60 s. Production problems may arise because the reaction mixture poured into the mold at the start may already have expanded, while the last poured fresh reaction mixture is still liquid. This can lead to over-rolling the fresh reaction mixture by the expanding foam, which may lead to the formation of foam defects or a poor density distribution along the board. To overcome this problem, the catalyst amount is steadily increased during the shot time. By doing so, the reactivity of the mixture continuously increases along the panel in the direction of the pour. As a result, the PU reaction mixture starts to cream simultaneously over the entire length of the mold, forming a uniform panel.

Continuous processing
PU laminated boards can be produced continuously using a double conveyor belt laminator. Fig. 7.19 shows a schematic representation of the laminator. Facings are fed into the laminator from below, the bottom facing, and above, the top facing. The facings fulfill a dual role: they prevent machine contamination and laminate the foam. The reactive mixture is poured onto the bottom facing, and when the foam has reached the end of its rise, it comes into contact with the top facing. The foam is cured in the heated laminator, and after sufficient cure, the continuous slab is cut to size and the panels are stacked. Fig. 7.20 on the right shows a fully operational laminator line, whereas the photograph on the left shows the laminator line under construction, providing a detailed view of the double conveyor belt.

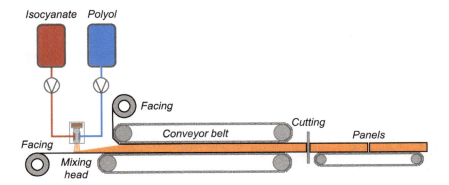

Fig. 7.19: Schematic representation of a double-belt conveyor for processing laminated boards.

The conveyor temperature for PU panel production is typically maintained at approximately 40 °C; however, for PIR foam, the laminator temperature is generally increased to 60 °C to facilitate sufficient isocyanurate formation at the surfaces. The A-component is prepared continuously by feeding the polyol and necessary additives from separate storage tanks into a premixing chamber. High-pressure metering pumps feed the A- and B-components to the high-pressure mixing head, where they

Fig. 7.20: Double conveyor belt for continuous production of laminated boards: left: under construction (with the kind permission of Cannon S.p.A.); right: completed and operational (with the kind permission of Hennecke GmbH).

are mixed. To ensure efficient mixing of polyol and polyisocyanate, the viscosity of the components must be relatively low (<5,000 mPa·s). The most commonly used physical blowing agent in Europe and Asia is pentane. Often, n-pentane is used for cost reasons. Pentane is generally directly fed into the mixing head.

The older liquid laydown technology uses a high-pressure mixing head attached to a bar, oscillating across the bottom facing. The reaction mixture is discharged via rakes for improved surface coverage. To achieve a proper laydown, the line speed of this technology is limited to about 20 m/min. Modern high-speed laminators with line speeds exceeding 60 m/min require different lay-down technologies. This technology uses fixed mixing heads equipped with special devices to distribute the reaction mixture across the width of the laminator [Fig. 7.21].

Fig. 7.21: Distribution devices used with fixed mixing heads (left: with the kind permission of Cannon S.p.A.; right: with the kind permission of BASF SE).

Insulation boards are produced with a thickness ranging from 20 to 200 mm, using various facing materials. The most common facings are thin aluminum foils, soda

Kraft paper, and coated glass mats. Insulation boards are predominantly used in construction applications such as wall, floor, and roof insulation [Fig. 7.22].

Fig. 7.22: Insulation board with aluminum facings (bottom rafter insulation) (with the kind permission of Covestro AG).

Fig. 7.23: Examples of profiled metal sandwich panels (with the kind permission of Covestro AG).

Steel or aluminum sheets are the common facing materials for metal sandwich panels. The facings are profiled to increase the rigidity of the panels [Fig. 7.23]. Profiling is achieved by passing the facers through fixed-mounted profiling rollers. Profiled metal sandwich panels are used for constructing industrial buildings.

7.5.3 Appliances

The manufacturing of appliances in one single operational step makes production highly efficient. The low thermal conductivity of PU foam enables the reduction of wall thickness, providing maximum usable space within the cabinet.

The cabinets are produced discontinuously using high-pressure mixing and day storage tanks. The physical blowing agent is pre-blended into the polyol component. Cyclopentane is the most used physical blowing agent in Europe. Combining cyclopentane with lower-boiling blowing agents, such as isopentane or isobutane, can increase the cell gas pressure. This allows foam density reduction at the expense of slightly higher thermal conductivities. HCFOs, optionally mixed with cyclopentane, are used in some cases to improve the insulation value of the foam. Due to their high output rates, L-shaped mixing heads can be used to minimize splashing.

Overpacking by at least 15% is required to fill the cabinets. The mold is preheated to approximately 40 – 45 °C to enhance curing and improve adhesion to the cabinet surfaces, thereby reducing the formation of dense skin. The maximum foam temperature is generally higher than the T_g of the foam. Therefore, the foam must cool in the mold below its T_g to prevent post-swelling of the cabinet. The dwell times increase with increasing wall thickness, requiring approximately 1 min per cm of foam thickness.

Fig. 7.24: (left): Schematic representation of the face-up pouring process; (right): multi-purpose mold that can be rotated to investigate different injection methods (with the kind permission of Covestro AG).

The preferred filling method is the so-called "face-up pouring" process [Fig. 7.24]. The appliance lies on its back, and the liquid reaction mixture is introduced from the side into the compressor step. The face-up method facilitates the uniform distribution of the foam, with short flow paths, and allows for the use of systems with increased reactivity. Advanced filling technologies were developed, such as "dynamic injection" (varying the output during the shot time) or "multipoint injection" (using several mixing heads to fill the cabinet).

The appliance industry typically measures the energy efficiency of the entire appliance. It accounts for foam imperfections, such as poor mold fill and density variations. The most common method is the "Reverse Heat Leakage" (RHL) test [9]. The cabinet with a heater inside is placed in a cold room and heated to a specified temperature. The energy required to maintain this temperature is reported.

7.5.4 Insulated pipes

The insulated pipes are prefabricated in factories and are ready for on-site installation. The space between the service pipe and the concentric protective outer pipe is filled with PU foam ("pipe-in-pipe insulation"). Insulated pipes are widely used to transport hot and cold fluids or gases [Fig. 7.25]. Depending on the diameter of the pipes, the PU layer can range from 5 to 250 mm. The PU foams require high shear strength to prevent foam damage during use, resulting from the thermal expansion of the pipes.

Fig. 7.25: District heating pipes insulated with PU (with the kind permission of Hennecke GmbH).

There are several production techniques to produce the pipes:
- Casting: The pipe assembly, closed on one end with a cap, is positioned at a slight angle, and the liquid reaction mixture is poured into the cavity between the service pipe and the outer pipe at the elevated end. Due to the long flow path spanning the entire length of the pipe, relatively large differences in density and compressive strength are observed along the pipe's length.
- Mixing head draw-through method: This technology eliminates the long flow path by drawing the mixing head through the cavity between the service pipe and the outer casing, thereby distributing the reaction mixture uniformly along the entire length of the pipe.
- Spray application: This method is primarily used for pipes with a large diameter. High-pressure machines spray the liquid PU reaction mixture over the rotating service pipe. Subsequently, the outer protective casing is applied, for example, by spraying a non-cellular PU top layer.
- Continuous casting: The inner service pipe is fed via a centering device into the production line, and the outer casing is formed from pliable metal or plastic foil. The foil is bent around the inner pipe, leaving space for injecting the PU foam.

7.5.5 Spray polyurethane foam

Spray Polyurethane Foam (SPF) is used for refurbishment and retrofitting [Fig. 7.26]. The seamless insulation layer provides thermal insulation and airtightness. The processing machinery for SPF foams is transportable and can be used on-site for pour-in-place. The reactivity of SPF systems must be high to prevent sagging. Spray foam formulations generally cure within seconds. These short reaction times can be achieved using polyols with high reactivity, such as Mannich base polyols.

Fig. 7.26: Insulation of a framed wall with PU spray foam (with the kind permission of SPFA (Spray Polyurethane Foam Alliance) and Graco Inc.).

The PU foams applied can be either open or closed-cell. Closed-cell SPF is typically applied at densities ranging from 24 to 32 kg/m^3. It is air and watertight and has a λ-value of about 24 mW/(m·K). Open-cell SPF exhibits inherently higher conductivity (35 – 40 mW/(m·K)) but can be applied at foam densities as low as 8 kg/m^3. Due to its open-cell structure, it is an effective material for sound absorption.

7.5.6 One-component foam

"One-component foam" or OCF refers to moisture-curing foams dispensed from pressurized cans. It is sometimes also used to describe self-curing two-component can foams. The more correct labeling of the second is "spray foam can" or SFC technology.

OCF is used in the building industry for applications such as window frame mounting, where the foam fills the gap between the wall and frame, fixing the frame to the wall [Fig. 7.27]. OCF is applied from the can via a plastic tube with a valve or pistol. The froth cures with environmental moisture, which may take several hours. OCF is easy to apply and widely used in professional and do-it-yourself (DIY) applications.

The can contains a mixture of an NCO-prepolymer with blowing agents and additives, such as surfactants, cell openers, catalysts, and flame retardants. The prepolymers with 5 – 7 wt.% NCO contents are based on low-functionality PMDI and mixtures of two and three-functional polyols with low hydroxyl values. Gaseous physical blowing agents, such as butane, propane, dimethyl ether, and carbon dioxide, are used for pressurization and foam expansion. Because the froth cures by reacting

Fig. 7.27: Application of one-component spray foam (with the kind permission of Soudal N.V.).

with air humidity, partial cell opening is required to ensure moisture reaches the core of the foam.

SFC cans contain the two reaction components in two separate chambers. The two-component formulations resemble standard polyether-based PU foams for construction applications. A mandrel penetrates the membrane that separates the two components within the can by pressing a button. The components are mixed by shaking the can. Once mixed, the pot life is short, and the foam has to be applied soon after activation.

References

Further reading

- G. Oertel, Polyurethane, Kunststoff Handbuch (Becker/Braun), Band 7, 1993, ISBN 3-446-16263-1.
- K. Dedecker, J. Deschaght, M. Barker in The Polyurethanes Book, D. Randall and S. Lee, Eds, Wiley, 2002, Chapter 15–17, ISBN 0-470-85041-8.
- D. Klempner, V. Sendijarevic, Handbook of polymeric foams and foam technology, Hanser Publications, Munich, 2004, ISBN 9781569903360.

Special references

[1] J.B. Emms, J.K. Gilham, Time-Temperature-Transformation (TTT) Cure Diagram: Modeling the Cure Behavior of Thermoset, J. Appl. Polym. Sci., 28 (8), 2567–2591, (1983).
[2] A.T. Huber, L.J. Gibson, Anisotropy of foams, J. Mater. Sci., 23 (1988) 3031–3040.
[3] P. Ferkl, M. Toulec, E. Laurini, S. Pricl, M. Fermeglia, S. Auffarth, B. Eling, V. Settels, J. Kosek, Multiscale modelling of heat transfer in polyurethane foams, Chem. Eng. Sci., 172 (2017) 323–334

158 —— 7 Rigid foams

[4] R.L. Glicksman, C.J. Hoogendoorn in Low density cellular plastics, N.C. Hilyard, A. Cunningham, Eds, Chapman and Hall, London (1994) ISBN 0 412 58410 7, Chapter 5 and 6.

[5] K.-E. Wagner, Simulation und Optimierung des Wärmedämmvermögens von PUR-Hartschaum, Dissertation, 2002, Universität Stuttgart.

[6] K. Dedecker, in The Polyurethanes Book, D. Randall and S. Lee, Eds, Wiley, 2002, Chapter 15, ISBN 0-470-85041-8.

[7] R. Leppkes, Polyurethanes, 6th Ed., 49–53, ISBN 978-3-86236-039-0.

[8] L.J. Gibson, M.F. Asby, Cellular Solids, 2nd Ed., 1997, Cambridge University Press, ISBN 0521-49560-1, Chapter 5.

[9] J. Sim, J. Ha, Experimental study of heat transfer characteristics for a refrigerator by using reverse heat loss method, Int. J. Heat Mass Transf., 38 (2011) 572–576.

8 Flexible foams

Flexible PU foams are soft and resilient open-cell low-density materials. They can easily be compressed and quickly recover upon stress release. Their high flexibility and resiliency make them excellent cushioning materials for automotive, furniture, and bedding applications. Flexible foams are used in domestic and office furniture, mattresses and pillows, car seats, and automotive interior trim such as headrests.

Flexible polyurethane foams are produced in a reaction between an isocyanate, polyol, and water. Both TDI and MDI are used as isocyanates. The polyols employed are predominantly polyether polyols, but polyester polyols are also used, especially in technical foam applications. The polyether polyols generally have a functionality of three. The urethane and urea reactions occur simultaneously, and the carbon dioxide generated in the reaction of isocyanate and water expands the foam. The amount of water employed determines the foam density, which is generally between 15 and 80 kg/m^3. The foams exhibit an open-cell structure with a high degree of interconnectivity. Upon compression, air can move freely within and escape from the foam.

The matrix material comprises a polymer network with an intermediate crosslink density and a block co-polymer structure with alternating soft and hard chain segments. The high-molar-mass polyol serves as the soft segment, whereas the polyurea block formed through the reaction of water and isocyanate constitutes the hard segment. The hard and soft blocks are connected through urethane bonds. The polyol and polyurea chain segments are thermodynamically incompatible, resulting in phase separation during the reaction, forming the soft and hard domains. The hard domains are stabilized through urea-urea hydrogen bonding interactions and fulfill a dual role. They both act as fillers and physical crosslinks, which contribute to the polymer's hardness and strength. Because branched monomers are used, the polymer is also chemically crosslinked. The chemical crosslinks provide additional stability, especially at high temperatures when the hard phases are softened. The soft domains provide low-temperature flexibility, resiliency, and recovery properties. The matrix polymer is soft and elastic, with a polymer modulus of less than 50 MPa. The low modulus of the polymer, the low foam density, and the high open cell content are fundamental to the comfort properties of the foam.

The difficulty of flexible foam production lies in the opening of the cells at the end of the foam expansion process. The cell opening is triggered by gelation, the onset of which can be tuned by the composition of the foam recipe. Because the foam recipe also determines the foam's physical properties, it is clear that preparing open-cell foam at the targeted density and with the required mechanical properties is a challenge. Developing foam formulations for a specific application may require significant trial-and-error formulation work; however, a general theoretical framework exists for processing and mechanical properties.

https://doi.org/10.1515/9783110744583-008

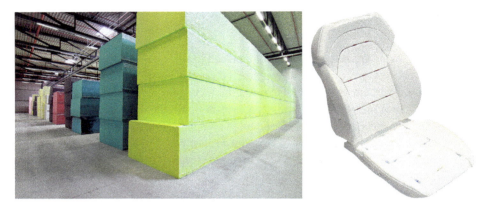

Fig. 8.1: (left) Slabstock-produced foam blocks; (right) molded automotive seating foam (with the kind permission of Hennecke GmbH).

Flexible foams can be produced continuously by slabstock foaming or discontinuously by foam molding. Using the slabstock process, large quantities of block-shaped foams are produced [Fig. 8.1; left]. The large foam blocks are then cut into smaller pieces for applications such as mattresses and upholstery. The molding process can produce foam in the desired shape, eliminating the need for cutting. Typical examples of molded foam products are pillows and automotive seats [Fig. 8.1; right].

Foams for a given application must fulfill several property requirements demanded by the foam fabricating industry, such as Automotive Original Equipment Manufacturers (OEM). Manufacturing foams with the specified properties can be challenging because they often require production using existing machinery and low system costs. The key factors for foam processing are:
- Manufacturing of defect-free and open-cell foam at targeted foam hardness and density.
- Flowability and foam stability properties during foam expansion.
- Cure/demolding time.

During foam rise, the foam must be sufficiently stable; in molding applications to allow for complete mold fill, and in slabstock, to prevent foam bun rupture. The production of open-cell foams requires the correct development of polymer strength at the onset of cell opening.

The main foam cost variables are:
- Foam density.
- Raw materials cost.
- Production cost.

8.1 Foam properties

Flexible foams must meet specific requirements depending on the application. The main requirements are the foam hardness, density, resilience, and airflow permeability. Furthermore, the foams need to fulfill basic mechanical strength properties such as tensile and tear strength. Compression set and dynamic properties may also be required. Flexible foams used in indoor applications must meet stringent standards for odor and emissions. Domestic and automotive foam applications may have to meet flame resistance criteria.

8.1.1 Basic mechanical properties

The foam density and hardness are basic application requirements, and foam resilience is a fundamental comfort factor. The foam density is measured per ISO 845. The compression hardness can be measured using a testing geometry where the compression plates are larger than the foam sample (ISO 3386) or by indentation of the foam with a smaller plate (ISO 2439). The ball rebound test (ISO 8307) is the simplest and most widely used method for measuring the resilience of a foam. The ball rebound resilience test begins by dropping a 16 mm-diameter steel ball from a height of 50 cm onto a foam test piece and measuring the rebound height. The rebound resilience is then calculated as a percentage of the rebound height relative to the drop height. High-resilience foams are preferred because they show shallower compression curves, resulting in a more even pressure distribution under load. Furthermore, they exhibit less hysteresis during a compression cycle, resulting in improved comfort and body support. The porosity and breathability of the foam also relate to comfort. The open-cell structure allows airflow through the foam, enabling the dissipation of heat and humidity away from the occupant. Airflow permeability (ISO 4638) is assessed by pulling air through a foam specimen at a constant pressure drop and is expressed in L/min. Moreover, the foam must meet the minimum requirements for basic mechanical properties, such as tensile strength and elongation at break (ISO 1798) and tear strength (ISO 8067). These properties are especially important when handling the foams in production, but are generally over-specified for final use.

8.1.2 Compression set and dynamic properties

The basic principle of compression set tests is to determine the residual deformation of a foam test specimen under prolonged compression at elevated temperatures over an extended period (ISO 2440). The compression set is the percentage of specimen deflection measured 30 min after decompression and recovery under ambient conditions. The temperatures, percent compression, and compression time are specified in the various test methods. The compression set measurement can be performed under both dry and humid conditions, or after autoclave aging (HACS, Humid Aged Compression Set). The most commonly applied accelerated creep test is performed at a compression of 50, 75, or 90% of the original specimen thickness at 70 °C for 22 h.

Dynamic tests evaluate the behavior of the foam under dynamic conditions over a specified period, indicating its useful service life. The most widely used fatigue test for furniture is the pounding test (ISO 3385) where the compression hardness and height loss are determined after 80,000 load cycles. For the automotive industry, dynamic creep and comfort tests have been developed. The essence of a dynamic creep test is that small sinusoidal load cycles are applied on a pre-compressed foam at a constant load. The changes in dynamic modulus and foam height are recorded with time, which measures the loss of comfort. Another requirement specific to automotive seating applications involves transmitting car vibration through the foam to the body. The seating foam provides comfort to the driver by decoupling the body from the car vibrations during the ride. Customer-specific tests were developed to determine the dynamic comfort of seating foams. The vibrations at low frequencies are amplified, but above 6 – 10 Hz, the foam acts as a damping element. The overall frequency amplification and dampening behavior, as well as their retention during the test period, depict the foam's ride comfort.

Polyurethane foams absorb noise in automotive applications such as roofliners and carpet underlay. The sound absorption coefficient of the foam can be determined using an impedance tube (ASTM C 384).

8.1.3 Flame resistance

Flammability tests have been developed for the automotive and furniture industries. These tests are essentially ignitability tests that use small open flame sources. To pass the fire tests, minimum flammability criteria must be achieved.

Crib 5 is a standard flammability test for upholstered furniture, as described in BS 5852. The test rig is a simulated chair comprising two flexible foam pillows (one horizontal and one vertical) covered with a fabric, and the ignition source is placed in the corner where the two foams meet at a right angle. The material must undergo three tests before certification, with Crib 5 being the final test. The first test is the cigarette test, in which a smoldering cigarette is applied to the material. The second is the match test, in which the equivalent of a burning match is applied to the material. The Crib 5 test uses a crib made of glued wooden planks containing a lint wetted with some drops of propylene glycol as an ignition source. The structure is placed on the upholstery material and set alight [Fig. 8.2]. This test is passed when neither the foam nor the fabric burns or smolders after 10 min.

The minimum flammability requirements for interior materials in passenger cars, trucks, and buses are specified in Federal Motor Vehicle Safety Standard No. 302 (FMVSS 302). A foam slab is aligned horizontally and exposed to a small Bunsen

Fig. 8.2: Crib 5 test for upholstered furniture (with the kind permission of SATRA Technology).

burner flame at one edge [Fig. 8.3]. The flame is applied for 15 s and then removed from the sample. The rate of flame spread across the sample is measured between two points. The maximum burn rate may not exceed 102 mm/min to pass the test.

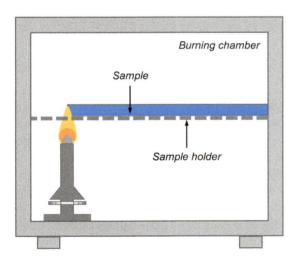

Fig. 8.3: FMVSS 302 test for automotive materials.

8.1.4 Emissions and odor

Emissions of flexible foam may be of concern because their open-cell structures allow prompt diffusion of non-bound organic substances from the polymer into the surroundings. OEM requirements for indoor air quality have made Volatile Organic Compounds (VOC) emission reduction increasingly important to the PU industry. Automotive manufacturers continually demand lower emissions (total VOC, aldehydes, amines, and aromatics) for vehicle interiors to minimize the potential health impact of volatile substances on car occupants and the typical new-car smell. The German VDA 278 method is the most widely used test for measuring foam emissions.

8.2 Flexible foam formulations

The basic formulation for flexible foam comprises the reactive components, polyol, water, and isocyanate. Catalysts and surfactants are always required, whereas the addition of crosslinkers is optional. The weight amount of the foam ingredients is expressed as amounts per 100 parts by weight of polyol [Tab. 8.1].

Tab. 8.1: Generic flexible foam recipe.

Component	Amount (p.b.w.)
Flexible polyol	100
Catalyst	1
Crosslinker	0–1
Surfactant	1
Water	3–5
TDI (index 100–110)	30–60
MDI (index 85–110)	50–80

Polyether and polyester polyols are used; however, almost all flexible foams are based on polyether polyols. Polyether polyols are produced from ethylene and propylene oxide using glycerine as a starter and potassium hydroxide as a catalyst. The equivalent mass of the polyols ranges from 1,000 to 2,000 g/mol. The catalysts can be tin- and amine-based. Tin-based catalysts are highly reactive and promote the urethane (gelling) reaction. Tertiary amines are less strong catalysts and less selective in promoting the urethane or urea (blowing) reaction. The most commonly applied amine catalysts are the gelling and blowing catalysts triethylene diamine (TEDA) and bis(2-dimethylaminoethyl) ether (BDMAEE), respectively. The crosslinkers can be diethanolamine, triethanolamine, glycerine, or short glycerine polyether polyols. The most common crosslinker, however, is diethanolamine. The surfactants are polydimethylsiloxane polyether graft copolymers. When using TDI, standard TDI is used, with a 2,4- to 2,6-isomer ratio of 80:20. MDI is used as a mixture of 2,4'-, 4,4'-MDI, and PMDI or prepolymers thereof. Because their NCO values differ (48 and 33 wt.% for TDI and MDI, respectively), the weight of isocyanate required to meet the targeted index is different for the two isocyanates. Mixtures of PMDI and TDI are also used. Depending on the majority compound, they are called MT or TM (T and M for TDI and MDI, respectively) blends.

Two fundamentally different types of polyether polyol are applied: slab and molded polyol. The two polyols differ in copolymer structure and molar mass. Slab polyols exhibit a random copolymer structure, whereas molded polyols show a blocked structure. Slab polyols are used in "conventional slabstock" foam and are always combined with TDI. Molded polyols are used in "High Resilience" (HR) foam and can be combined with MDI and TDI. HR polyols are used in the production of both molded and slabstock foam.

Slab polyols are produced using a random feed of ethylene and propylene oxide with approximately 5 – 15% ethylene oxide. Because ethylene oxide is more reactive than propylene oxide, the ethylene oxide concentration will gradually decrease during the reaction, and propylene oxide will likely react last, giving a secondary alcohol as the chain terminal group. The secondary hydroxyl content of slab polyols is generally higher than 97%. The primary function of the ethylene oxide units in the polyol is to enhance compatibility with the other reaction components, particularly water. These polyols typically have an equivalent mass of 1,000 and a molar mass of 3,000 g/mol.

Molded polyols are block copolymers of ethylene and propylene oxide. Ethylene oxide is added after the propylene oxide addition reaction is completed. The ethylene oxide content in these polyols ranges from 10 to 20%. The end-capping with ethylene oxide provides the polyol with primary hydroxyl groups, the content of which increases with increasing amounts of ethylene oxide. The molded polyol primary hydroxyl group content ranges from 75 to 85%. The primary function of the polyethylene oxide cap is to enhance the reactivity of the polyols, which are therefore also referred to as "reactive polyols". When using molded polyols, the urethane reaction is generally faster than the urea reaction. The rapid urethane reaction results in an early increase in viscosity, which enhances stability during the foam rise. Furthermore, the ethylene oxide cap increases the compatibility with water and isocyanate. The polyol equivalent and molar mass of molded polyols range from 1,500 to 2,000 g/mol and 4,500 to 6,000 g/mol, respectively.

The compression hardness of foam at a given foam density can be selectively increased by incorporating filled polyols, such as graft or polymer polyols, PIPA (Poly-Isocyanate Poly-Addition), and PHD (Poly-Harnstoff-Dispersion) polyols. The carrier polyol can be a slab or a molded polyol, allowing filled polyols to be used in conventional slabstock and molded foam applications.

Flexible foams are expanded with carbon dioxide. The prime source of carbon dioxide is the reaction between isocyanate and water, but it can also be applied as a physical blowing agent. Using carbon dioxide as a physical blowing agent is an established technique in slabstock and molding. Gaseous carbon dioxide is injected directly into the polyol or isocyanate line before it enters the mixing head. Carbon dioxide-assisted foaming is well established and can be carried out with relatively minor machinery modifications.

8.3 Polymer topology and morphology

The polyaddition reaction involves three components: polyol, water, and isocyanate. The employed polyols are three-functional, and crosslinkers and branched isocyanates are optionally used. The reaction yields a crosslinked multi-block-copolymer structure with alternating soft and hard blocks. The polymer exhibits a phase-separated structure due to the thermodynamic incompatibility between the hard and soft blocks. A nanometer-sized microphase morphology is formed, and round-shaped micron-sized aggregates can be formed. The microphase morphology is obtained through a reaction-induced spinodal decomposition mechanism. The hard domains exhibit an interconnected lamellar morphology, giving high hardness at relatively low hard block contents. The soft segments give low-temperature flexibility.

8.3.1 Crosslinked block-copolymer structure

Let us consider a typical foam recipe, as shown in Tab. 8.1. The polyol has an equivalent mass (EM) of 2,000 g/mol; the blend comprises 3.6 parts water, and TDI is used as isocyanate at an index of 100. The urethane-to-urea ratio, the hard segment content, and the average chain topology of the polymer are calculated as follows.

A sample of 100 g of the polyol contains 0.05 mol of hydroxyl groups that will be converted to urethane in the reaction with the isocyanate. For every water molecule ($M = 18$ g/mol), one urea group will be formed; thus, 0.2 mol of urea will be produced. Accordingly, the molar ratio of urethane to urea amounts to 1:4. The number of urea groups in the hard block is then twice this number, hence 8. Fig. 8.4 illustrates the chain topology between two polyol crosslinks, where a urea hard block is connected to polyol branches on both sides. It schematically represents the "average" polymer topology; however, in reality, the mass distribution of the hard blocks is broad.

A simplified calculation of the hard block content (HBC) is proposed that only includes the weights of polyol and isocyanate, ignoring the weight of water and the weight loss due to the release of carbon dioxide.

$$HBC = 100 \cdot \frac{isocyanate\ weight}{isocyanate\ weight + polyol\ weight}$$

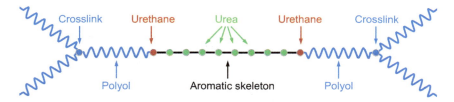

Fig. 8.4: A schematic representation of the urea hard block between two polyol crosslinks.

The polyol blend, comprising polyol (EM = 2,000 g/mol) and water (EM = 9 g/mol), contains 0.45 equivalents of isocyanate-reactive groups. The weight of TDI (EM = 87 g/mol) required to achieve an index of 100 equals 39 g. Accordingly, the HBC amounts to 28 wt.%.

The soft block, predominantly based on propylene oxide, is apolar. The hard block is polar because of its high urea content. The polarity difference causes hard and soft blocks to phase-separate, forming hard and soft domains. The hard domains are stabilized through interchain interactions, the urea-urea hydrogen bonding interaction being the strongest. However, the ordering of the hard blocks in the hard domains is poor because of geometrical constraints that prevent close packing.

The poor stacking is related to the isocyanate composition and the optional use of crosslinker polyols. MDI is always employed as a mixture of 2,4'-, 4,4'-MDI, and PMDI and TDI as an 80:20 blend of 2,4- and 2,6-TDI. Due to their low molar masses, crosslinkers form crosslinks within the hard domains. Mixed urea structures and hard-phase crosslinks hinder the alignment of hard blocks. The hard domains are amorphous and show rigid glass-like properties.

The soft phase is crosslinked because the polyols are three-functional, and when crosslinkers or branched isocyanates are used, the hard phase will also be crosslinked. The overall crosslink density of the polymer is, therefore, substantial. For most flexible foams, the average molecular mass between crosslinks ranges from 1,000 to 5,000 g/mol. The excellent long-term recovery properties and high-temperature stability of flexible foams can largely be ascribed to the chemical crosslinks.

8.3.2 Two-scale morphology

Fig. 8.5 shows AFM images of a conventional TDI foam strut produced with a relatively high water content. The magnification increases from A to C. Increasing local hardness changes the arbitrary color scale from dark brown to light brown. The light and dark phases are rich in polyurea and polyol, respectively. The AFM images are two-dimensional projections of a three-dimensional structure; therefore, some care must be taken in their interpretation.

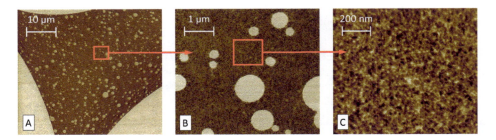

Fig. 8.5: AFM recordings of the cross-section of a flexible foam strut at three levels of magnification. The samples were microtomed and exhibited a smooth surface (with the kind permission of BASF SE).

At low magnification, the contours of the strut embedded in epoxy resin are visible [Fig. 8.5 A]. At a ten-fold higher magnification [Fig. 8.5 B], round macrophases with diameters of 0.1 – 1 µm can be observed, which are substantially harder than the surrounding material. These macrophases are known as urea "balls", "aggregates", or "precipitates". Although the aggregates contain predominantly urea, some polyol is incorporated. Their urea content was measured to be 2–4 times that of the surrounding polymer. The formation of the macrophases is system-dependent. For instance, they are observed in systems based on slab polyol but not in formulations based on molded polyol [1].

The occurrence of aggregates is related to the water compatibility of the polyol and the water content in the formulation. Apolar polyols and high water contents favor the formation of the macrophases. Aggregates are formed when the concentration of water exceeds the solubility of water in the reacting mixture. The water droplets react with isocyanate, provoking locally high urea oligomer concentrations that phase separate and form the aggregates. Their round shapes indicate that their formation occurs early during the reaction when the viscosity of the reacting liquid is still low.

The highest magnification [Fig. 8.5 C] reveals the micro-segregated phase morphology of the continuous polymer matrix. The hard domains fill space relatively uniformly, appear geometrically anisotropic, and are reasonably well segregated from the soft phases. The center-to-center spacing of the hard domains is approximately 10 nm. Two-thirds of this distance accounts for the polyol layer, and the thickness of the hard domains accounts for approximately one-third. The thickness of the soft layer is determined by the molecular mass or, more precisely, by the radius of gyration of the polyol. During the reaction, the hard blocks in the block-copolymer structure can drift apart during the phase separation process, but the gap between them is determined by the distance the polyol can bridge. The thickness of the hard domain is, in essence, determined by the hard block content. More detailed studies revealed that the hard domains exhibit lamellar-type structures with dimensions of approximately 10 – 20 nm in width and 50 – 100 nm in length. The hard domains appear interconnected, forming a bicontinuous structure. The formation of this microphase morphology originates from a reaction-induced spinodal decomposition mechanism [2].

Spinodal decomposition is a phase separation mechanism by which a single homogeneous phase spontaneously separates into two phases. Phase separation happens when the homogeneous phase becomes thermodynamically unstable. This is what occurs during the foaming reaction. After mixing the reactants, the urethane and urea reactions start. The polyol gets capped with isocyanate, and the chain length of the urea oligomers increases. As a result, the interaction between the polyether and polyurea segments decreases as the conversion of the isocyanate increases. The initial homogeneous phase becomes unstable once the urea segment length exceeds a critical value and microphase separation starts. Soon after that, hydrogen-bonding interaction occurs between the urea groups. The polyol-rich and urea-rich phases develop uniformly throughout the volume, showing an intertwined morphology with a lamellar ordering. The contours of the final phase morphology are already mapped out at the onset of phase separation with a length scale set by the polyol gyration radius. With continuing reaction, the phases grow in amplitude. Further urea formation and urea hydrogen bonding interaction lead to vitrification of the hard phases. Vitrification quenches the microphase formation process, thereby increasing the polymer strength. The obtained morphology comprises an interconnecting physical network of hydrogen-bonded urea hard blocks within the crosslinked polyether-urethane [2, 3].

8.4 Manufacturing of open-cell foam

A critical step in manufacturing flexible foams is the opening of the cells near the end of the foam rise. The cell opening is triggered by polymer gelation. The gelation is of physical origin and occurs when the hard domains start to vitrify. The mechanical strength the polymer has attained at the cell opening determines the stability of the rising foam, which, in turn, determines the density, and cell porosity of the foam.

8.4.1 The cell opening mechanism

The foam producer observes the cell opening of a flexible foam as a "blow-off" – the sudden cessation of foam expansion, characterized by the bursting of surface bubbles and the release of gas under pressure. When cell opening occurs, foam expansion stops abruptly, and the foam sighs back somewhat. It was already concluded in the early days [4] that the sudden gelling or stiffening of the polymer triggered the cell opening. The sudden stiffening, however, occurs well before chemical gelation. Hence, the polymer exhibits a modulus but lacks mechanical strength in the absence of a covalent network. Consequently, the cell membranes cannot accommodate the strain resulting from the continued expansion of the cells; the cell membranes rupture, and the cell opening process begins.

For example, in a series of TDI slabstock foam formulations with an increasing water content of 2 to 6 parts per 100 parts by weight of polyol, the calculated conversion at chemical gelation increased from 84 to 90% as the water content increased. The experimentally determined conversion of bubble bursting, however, occurred well before chemical gelation and decreased with increasing water content from 82 to 50% [5].

The understanding of the foaming and cell-opening process was further refined by studying cell opening through a combination of real-time IR and SAXS, as well as rheology [3]. The MDI-based recipe employed in this study exhibited reduced reactivity and maintained stable foam formation until cell opening. SAXS and IR measurements were used to determine the kinetics of phase separation. Four stages of foam rheological development were observed: *i)* mixing and bubble nucleation, *ii)* liquid foam expansion and microphase separation, *iii)* physical gelation triggering cell opening, and *iv)* finalization of the flexible foam structure [3, 6].

The first stage *(i)* commences with mixing and initial bubble formation and ends when the bubble network is formed, which occurs at a conversion of about 30%. With continued reaction *(ii)* and generation of carbon dioxide, the foam further expands. An increase in scattering intensity was observed in SAXS at an isocyanate conversion of 52%, marking the onset of the microphase separation transition. The scattering intensity and urea hydrogen bonding increased rapidly after the onset of phase separation, indicating the development of the hard phases due to the ongoing formation of urea. The molar mass and viscosity of the reacting liquid at this stage were still low, and the foam elastic storage modulus (G') showed a broad plateau value till about 70% conversion. Having reached this conversion, the phase separation process was intercepted by hard phase vitrification; the hard phase reached a T_g equal to the reaction temperature. Vitrification of the hard domains freezes the morphology, and the rapid increase in scattering intensity and urea hydrogen bond interaction slowed down. Macroscopically, the foam had reached its gel point, and cell opening occurred, marking the end of stage two. During the third stage *(iii)*, there is a significant increase in the G' of the foam, approximately two orders of magnitude, resulting from the continued vitrification of the hard phase and concomitant hardening of the polymer. During the fourth and final stage *(iv)*, the reaction continues, albeit at a much slower rate, due to diffusion control and the depletion of reactive groups. The foam reaches its maximum temperature with time and full conversion.

It is important to note that the morphological development and final morphology of MDI- and TDI-based systems are similar, regardless of the polyol employed.

8.4.2 Foam stability

The cells are closed during foam expansion, and the gas pressure generated by the carbon dioxide produced in the urea reaction expands the foam. At the cell opening, the gas escapes, the pressure in the foam drops, and the expansion stops. To endure the cell opening event, the struts must be strong enough to bear the weight of the foam. When the gas escapes, the foam sighs back, typically between 1 and 10% of its maximum height [Fig. 8.6b]. The continued reaction strengthens the foam; after 1 – 3 min, the foam is strong enough to be handled.

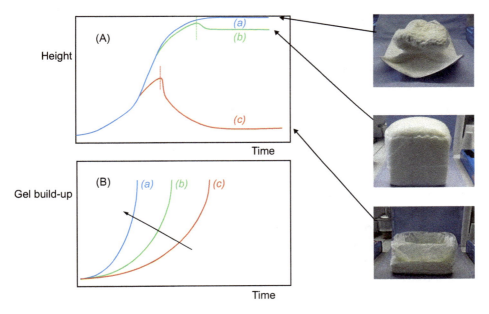

Fig. 8.6: The height (A) and gel build-up (B) versus time. Formation of: (a) tight foam showing severe shrinkage, (b) stable and open-cell foam, (c) unstable foam giving foam collapse. The gel profiles correspond to the height profiles in the same color (a, b, and c). The dotted line indicates the onset of cell opening and is absent in (a), the black arrow indicates increased gelling activity (foam pictures with the kind permission of BASF SE).

The conversion at which cell opening occurs determines the density and airflow permeability of the final foam. When cell opening occurs prematurely, and the struts have not attained sufficient strength to cope with the pressure drop, the foam at cell opening will be unstable. With the decreasing strength and stability of the polymer struts at the cell opening, the foam may exhibit strong recession with increased foam density, or, in more severe cases, foam split or even foam collapse [Fig. 8.6c]. In contrast, when cell opening is delayed, the cell opening may be incomplete. The foam can be "tight" and exhibit poor airflow upon deformation (pneumatic foam) or

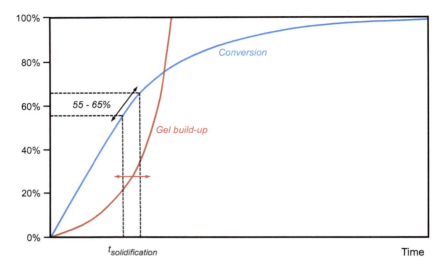

Fig. 8.7: Isocyanate conversion (%) and gel/molar mass build-up (arbitrary units) as a function of time. The arrows indicate the possibility of changing the onset of physical gelation (black) and the molar mass build-up (red).

may exhibit severe shrinkage at high closed cell contents. Foam shrinkage is the combined result of cooling hot gas and gas diffusion. Carbon dioxide diffuses out much faster than the air coming in, causing under-pressure in the cells [Fig. 8.6a].

Cell opening occurs when physical gelation begins. This occurred at an isocyanate conversion between 50 and 85% in the series of TDI foams with increasing water content, as previously discussed [5]. In Fig. 8.7, the spread is narrowed to 55 and 65%, a typical conversion range for cell opening of molded foams. The strength of the polymer is related to the gel build-up, which is determined by the polymer molar mass and urea hydrogen bond interaction. The molar mass is determined by the functionality of the starting monomers, the isocyanate conversion, and the ratio between the gelling and blowing reactions. Urea hydrogen bonding depends on the amount of urea formed and the geometrical structure of the urea hard block.

The foam stability at cell opening can be adjusted by modifying the recipe. The red and black double arrows in Fig. 8.7 indicate how, in principle, the gel profile and conversion at gelation can be adjusted [7]. In foam recipes based on molded polyol, the gel profile, and consequently the polymer strength development, can be adjusted by choice of catalyst and polyol:

- Catalyst adjustment: The polymer molar mass at a given conversion of isocyanate can be steered by manipulating the gel-to-blow ratio. Increasing the gel-to-blow catalyst ratio enhances the gel build-up, causing the gel build-up curve in Fig. 8.7 to become steeper. This leads to increased polymer strength at cell opening and concomitantly improved foam stability. However, the foam can become over-stabilized when the gelling reaction is promoted too strongly. The increased molar mass makes the cell membranes more elastic, which may prohibit membrane rupture. As a result, the foam may be tight. In contrast, a decreased gel-to-blow ratio slows the gel build-up, and the gel build-up curve becomes shallower. The foam stability is reduced, which may result in an earlier cell opening and lead to an increase in foam density. When adjusting the catalyst ratio, care must be taken to ensure that the overall reactivity remains constant, allowing for a fair comparison. Increased reaction rates delay the phase separation process, improving foam stability and *vice versa.*

– Functionality and reactivity: The gel profile can also be adapted by altering the functionality and intrinsic reactivity of the polyol. The gel build-up develops more rapidly with increasing functionality and polyol reactivity, thereby improving foam stability. The functionality and reactivity can be changed in the following manner. Flexible foam polyols are generally three-functional, but some polyols with higher and lower functionality are available to adjust the average functionality of the polyol blend. The reactivity of molded polyols corresponds with their primary hydroxyl content, which is determined by the ethylene oxide cap length. Hence, the use of polyols with increased EO cap length will increase foam stability at cell opening and *vice-versa.*

The conversion at physical gelation and concomitantly cell opening can be altered by index variation and isocyanate composition:

– Index variation: As the index increases, the mass fraction of isocyanate in the system also increases. An increase in isocyanate content at a given isocyanate conversion results in the formation of more and longer urea oligomers. Their solubility in the reactive mixture is reduced, and the onset of phase separation and physical gelation occurs at lower conversions. The molar mass of the polymer at the cell opening is reduced, resulting in decreased foam stability. A reduction in the isocyanate index shows the reverse effect; the foam stability improves.

– Variation of isomer ratio: The standard isocyanate for flexible foam production is T80, a mixture of 2,4- and 2,6-TDI at a ratio of 80:20. Foams also can be prepared from a 50:50 mixture of 2,4'- and 4,4'-MDI (without PMDI). The isocyanates 2,6-TDI and 4,4'-MDI are symmetrical, whereas 2,4-TDI and 2,4'-MDI are asymmetrical. Urea chain segments from symmetrical isocyanates form stronger intermolecular interactions (e.g., formation of hydrogen bonds) than their asymmetrical counterparts [Chapter 9.3.4]. Thus, when the symmetrical isocyanate content in the isomer mixture is increased, physical gelation and cell opening will occur at lower conversions, leading to reduced foam stability and *vice versa.*

8.5 Morphology and polymer hardness

The spinodal phase separation process forms a bicontinuous structure of hard and soft domains characterized by lamellar ordering at relatively low hard block mass fractions. The hard domains exhibit rigid/glassy properties and are sandwiched between low-modulus soft domains. The block-copolymer structure ensures a perfect bond between the two phases, providing continuity of displacement across the phase boundary. The hard domain co-continuity provides the flexible foam polymer with a significantly higher polymer modulus than a discontinuous particle-like hard phase morphology would have given.

Davies [8, 9] developed an expression to predict the shear modulus of materials with two interpenetrating continuous phases, where φ_A and G_A represent the volume fraction and modulus of phase A, and similarly for φ_B and G_B for phase B:

$$G^{\frac{1}{5}} = \varphi_A \cdot G_A^{\frac{1}{5}} + \varphi_B \cdot G_B^{\frac{1}{5}}$$

The modulus of the soft and hard phases is about 1 MPa and 1 GPa, respectively. The three-order magnitude difference between the two moduli implies that the soft phase contributes little to the shear modulus. Thus, the equation can be rewritten as:

$$G = G_{HPh} \cdot \varphi_{HPh}^{5}$$

with G_{HPh} and φ_{HPh}, the modulus and mass fraction of the hard phase, respectively. Experimentally, this dependency of the foam hardness on the hard block content could be confirmed; the polymer modulus scaled with about the fifth power of the hard block content.

8.5.1 Modulus-temperature behavior

The microphase morphology of hard and soft domains is reflected in the modulus temperature behavior of flexible foam. Fig. 8.8 illustrates the DMA results of a high-resilience MDI-based flexible foam. It displays the storage modulus (G') and loss factor (*tan δ*) versus temperature. G' and the loss factor relate to the foam hardness and resilience, respectively. The polymer exhibits glass-like properties at low temperatures; its molecular mobility is highly restricted, resulting in a hard and brittle foam. The foam starts to soften at approximately -60 °C, marking the onset of the soft phase glass transition. The glass transition is broad and reaches 0 °C. Passing the glass transition, G' declines over almost two orders of magnitude. Above 0 °C, the sharp decline in modulus slows down, and the further decrease in G' with temperature is only modest. The G' value of the rubbery plateau is related to the mobility of the soft blocks, which reduces with increasing hard block content and, to a lesser extent, with increasing crosslink density of the polymer.

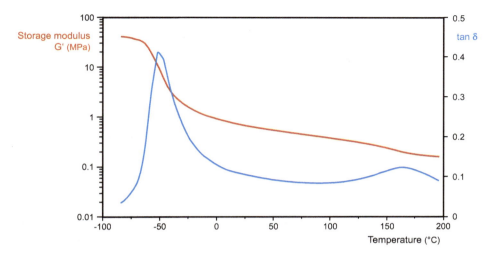

Fig. 8.8: Modulus G' and loss factor (*tan δ*) versus temperature of an MDI-based high-resilience flexible foam, recorded at a frequency of 1 Hz (with the kind permission of BASF SE).

Complementary information on the phase behavior can be derived from the loss factor. Its value increases and reaches a maximum during a phase transition, occurring at approximately -50 and 170 °C for the soft and hard phases, respectively. The maximum value in the loss factor at approximately -50 °C coincides with a sharp drop in G'. The high value of the loss factor indicates that the foam at and around this temperature is highly damping. The loss factor decreases with a further increase in temperature, implying that the foam becomes less damping and gradually more resilient. The width of the *tan δ* peak of the soft phase transition suggests that the glass transition is broader than concluded from the G' versus temperature curve. The *tan δ* curve clearly shows a tail extending into the room temperature region, indicating that the resilience of the foam improves steadily with increasing temperature within the range of 0 – 50 °C.

When the molar mass of the polyol is increased, the soft domains become thicker, and the phase separation between hard and soft domains is enhanced. The G' curve shifts somewhat to lower temperature, and the drop in G' is steeper. The *tan δ* peak is also shifted to lower temperatures, is narrower, and has a higher maximum value. The values of G' and the loss factor at room temperature are reduced, indicating that the foam becomes softer and more resilient with increasing molar mass of the polyol.

The equivalent mass of molded polyols is approximately twice that of slab polyols. Consequently, foams prepared from molded polyols show a higher ball rebound resilience than slabstock foams. The former shows ball rebound values exceeding 50%, and the latter ranges between 30 and 40%.

The softening of the hard domains occurs at high temperatures. The G' curve exhibits a minor transition at approximately 150 °C, whereas *tan δ* shows a maximum at

170 °C. The softening of the hard phase shows little effect on the course of G'. When the physical crosslinks soften, the chemical crosslinks take over, and the foam largely maintains its structural integrity.

8.6 Compression hardness

The sample is conditioned by compressing it three times to 70%, and then the complete compression-decompression cycle is taken as the fourth run. The ratio of the areas of the loading and unloading curves (in percent) represents the energy loss during the cycle.

Fig. 8.9 shows the Compression Load Deflection (CLD) curve of a high-resilience flexible foam, as specified in ISO 3386. The stress-strain curve shows linear elastic behavior till a compression of about 5%. From there, it shows a plateau that transitions into a regime where the stress steeply increases. The first linear elastic part is governed by cell wall bending. The initial slope of the curve provides the Young's modulus (E^*) of the foam. At about 5% compression, the struts start buckling, and the stress increase subsides. The plateau corresponds to the collapse of the cells, where the struts bend and elastic hinges form. When most cells have collapsed under high compression, the

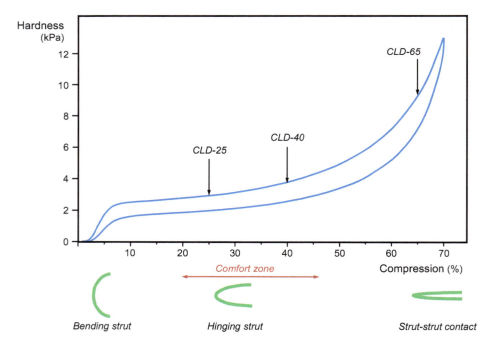

Fig. 8.9: Compression load-deflection curve (according to ISO 3386) of high-resilience foam at a density of 45 kg/m^3 (with the kind permission of BASF SE).

struts start to touch each other. Increased strain will lead to more strut-strut contact, and the stress rises steeply.

The foam hardness for a given application is typically selected so that it is compressed by approximately 25 – 50% of its original height under a specified load. Then, there is sufficient surface contact between, e.g., body and foam, and the struts are "hinging", both of which are important for a comfortable feeling. The CLD value at 40% compression is generally considered the foam hardness. However, the CLD values at 25 and 65% may also be recorded. The ratio of the CLD value at 65 and 25% is taken as the SAG or support factor of the foam. The SAG factor describes the relationship between sinking in and the supporting force, indicating the cushioning quality. A foam with a SAG factor between 2.5 and 3 should provide high comfort. The CLD-40 hardness of foams with densities ranging from 20 to 80 kg/m^3 may vary between 1 and 20 kPa.

8.6.1 Effect of density and polymer modulus

The foam hardness can be altered by changing the polymer hardness and the density of the foam. The (elastic) Young's modulus of the foam (E^*) correlates with the elastic modulus of the polymer (E_s) and the foam density (ρ) in the following manner [10]:

$$E^* \approx E_s \cdot \rho^2$$

The density of a foam produced from a given formulation can be decreased by using a physical blowing agent, such as carbon dioxide. In molding applications, the density can be increased by increasing the overpack factor. The CLD-40 value of two foams based on the same formulation but produced at different foam densities relate as given below:

$$\frac{CLD\text{-}40_{\rho1}}{CLD\text{-}40_{\rho2}} \approx \left(\frac{\rho_1}{\rho_2}\right)^2$$

where $(CLD\text{-}40)_{\rho1}$ and $(CLD\text{-}40)_{\rho2}$ are the CLD-40 values of the foams at densities ρ_1 and ρ_2, respectively.

E_s can be increased by adding fillers or increasing the isocyanate index. The most convenient way to introduce filler particles is by substituting a portion of the polyol in the formulation with polymer-polyol. Fig. 8.10 illustrates the effects of increasing the hard block content (A) and adding filler (B) at the same foam density.

An increased hard block content increases E_s, and the foam becomes harder. However, the T_g of the soft phase increases, and the phase transition broadens. Concomitantly, the loss factor at room temperature increases, resulting in the foam becoming less resilient.

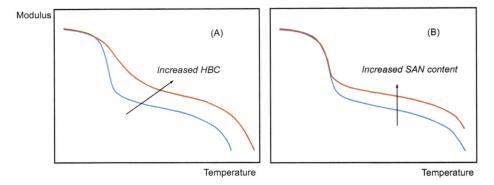

Fig. 8.10: Modulus versus temperature at constant density. The modulus increases with an increase in hard block content (A) and the addition of graft polyol (B).

The copolymer particles in graft polyols are rigid and show a T_g and Young's modulus of about 120 °C and about 1 GPa. Adding the particles increases the foam hardness but does not affect the glass transitions of the soft and hard phases of the polymer matrix material. Hence, the foam resilience remains the same. The production of high-resilience foam with high hardness, for example, for automotive seating, generally requires a graft polyol.

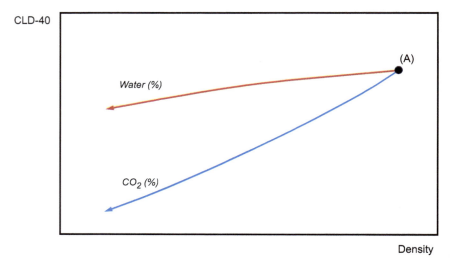

Fig. 8.11: Reduction in foam hardness by the addition of a physical blowing agent (CO_2 %) and an increase in water content (water %), starting from reference foam (A).

A special case is the increase in water content at a constant index. The hard block content increases, and so does E_s. However, the foam density is reduced simultaneously, which decreases the hardness. The effect of a physical blowing agent and an increase in water content on the hardness of the foam relative to its density is compared in Fig. 8.11, starting from a reference formulation, foam "A".

The physical blowing agent reduces the foam density, whereas the formed polymer remains unchanged. The foam hardness decreases with ρ^2, and the foam resilience is unaffected. An increase in water content requires an increase in the amount of isocyanate to maintain a constant index. The increased amount of gas reduces the foam density, whereas the increase in the hard block content increases E_s. The increase in E_s partially compensates for the reduction in density, resulting in a modest overall decrease in hardness.

8.7 Ball rebound resilience, hysteresis, and loss factor

The energy loss processes in flexible foam are governed by its chemical composition, polymer morphology, and cellular structure. The measured energy loss, however, also depends on the time scale and the extent of deformation during the test. There are three methods for determining the energy loss: the hysteresis loss during a compression cycle, the loss factor by DMA, and the ball rebound test. Their deformation frequencies are on the order of several seconds, one second, and milliseconds, respectively. Increased foam resilience is indicated by an increase in ball rebound and a decrease in hysteresis and loss factor. The resilience determined by the ball rebound test has the most practical relevance and has been correlated to the overall cushion comfort of foams.

The resilience of foam is, in principle, a polymer property; however, the airflow permeability of the foam may affect the measured resilience value [11]. The airflow permeability does not affect the loss angle, because the deformation in the DMA test is minimal. The foam deformation is substantial, and the contact time is short in the ball rebound test, and the foam openness may influence the measured rebound value. Foams with constricted airflow permeability ("tight" foams) can exhibit pneumatic damping, reducing the ball rebound. When the airflow rate (according to ISO 4638) exceeds 60 L/min, the ball rebound resilience is independent of the foam's airflow permeability. Being a polymer property, the resilience of a foam is independent of the foam density. For example, when the density of a foam is reduced by using a physical blowing agent, the ball rebound is not affected, provided that the airflow permeability of the foam remains sufficient.

8.8 MDI versus TDI technology

Flexible foams can be produced with TDI and MDI. TDI foam can be produced at lower densities and exhibits superior mechanical properties, whereas MDI foam production offers processing advantages.

The fundamental difference in properties of TDI and MDI-based foams lies primarily in the different NCO values and functionalities of the two isocyanates. The aromatic ring in TDI carries two isocyanate groups, whereas each phenyl ring in MDI has only one. This results in different NCO values for TDI and MDI, being 48 and 33%, respectively. Thus, a given A-component requires a higher mass amount of MDI than TDI to achieve the same isocyanate index [Fig. 8.12].

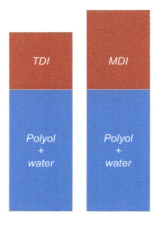

Fig. 8.12: The weight amounts of TDI and MDI required to achieve the same index for a given polyol blend.

This has two consequences. The TDI-based foam will show a lower density and a lower hard block content. A lower foam density is obtained because less material needs to be expanded with the same amount of carbon dioxide generated in the urea reaction. The lower hard block content gives a lower E_s and a softer but more resilient foam.

TDI, being two-functional, cannot form crosslinks. The MDI used in flexible foam technology contains PMDI and exhibits an average functionality between 2.15 and 2.3. Consequently, MDI-based foams show a higher degree of crosslinking than TDI foams. Crosslinks, however, reduce the extensibility of the polymer network and, concomitantly, the elongation and tensile strength at the break of the foam. Hence, TDI foams generally exhibit superior tensile and tear properties.

All isocyanates are respiratory irritants and potential sensitizers. The main hazard arises from inhaling vapors. MDI exhibits a 5,000-fold lower vapor pressure than TDI ($6 \cdot 10^{-4}$ Pa and 3.3 Pa at 25 °C, respectively). Therefore, MDI is easier to handle and requires lower industrial hygiene levels during PU production.

When using MDI, there are two aromatic rings per urea, whereas TDI has only one aromatic ring. Therefore, the hard domains in MDI are less polar. Consequently, the MDI-based hard domains will take up less moisture and soften less under humid conditions. As a result, MDI-based foams exhibit improved recovery in humid aged and wet compression set tests.

TDI is kinetically highly asymmetric. The first NCO group on TDI reacts readily, but the second is slow. TDI-based systems require strong tin-based catalysts or relatively high amounts of milder amine catalysts to cure the foam. MDI is more reactive overall and requires less catalyst, which reduces catalyst emissions. Emission reduction is especially important in automotive applications.

MDI builds up molar mass quickly because of its higher reactivity and functionality. This improves foam stability and widens the index latitude. The index can be varied broadly around 100, with little to no effect on foam stability. This enables the manufacture of dual-hardness automotive seats using one A-component. The soft base part of the seat is produced at a lower index, whereas the side bolsters are produced at a higher index. In contrast, TDI systems show a narrow index latitude. Using TDI, two different polyol formulations are required to produce the soft and hard parts, which requires more complex dispensing machinery.

8.9 Processing

Flexible foam can be produced by molding or slabstock foaming. Slabstock foaming allows the continuous production of large quantities of flexible foam. The large foam buns produced are cut to size for the final application. Molding is a discontinuous foaming process that produces the foam in the desired shape. The market for polyurethane flexible foam is divided by the foam production method, whereby the market share of slabstock foam exceeds that of molded foam.

8.9.1 Molded foam

Molded foams are used in several automotive applications. Seating is the biggest application, followed by foam-backed automotive carpets and headrests. Furthermore, molded flexible foam is used in various domestic furniture applications, including seats and cushions, as well as in orthopedic foam products and office furniture [Fig. 8.13].

Both high-pressure and low-pressure dispensing units are used to produce molded flexible foam. The required amount of reactive mixture is poured into an open or injected into a closed mold, and after the reaction is completed, the part is removed. A distinction must be made between the hot-cure and cold-cure processes [12].

Most molded foams are produced using cold-cure technology, where the mold temperature is typically maintained between 35 and 55 °C. Some preheating of the molds is required to ensure a uniform skin density and completion of the reaction of the foam layer in contact with the mold. Molded polyols are employed because they

Fig. 8.13: Various molded foam applications (left: with the kind permission of Krauss Maffei Technologies GmbH; right: with the kind permission of Ottobock SE & Co. KGaA).

allow low mold temperatures and fast demolding. TDI, MDI, and TDI/MDI mixtures are used as isocyanates.

The hot-cure process uses less reactive polyols with a short ethylene oxide-based cap and TDI as the isocyanate. The reaction requires mold temperatures over 150 °C to achieve full cure. After the reaction is complete, the mold must be cooled to 35 – 45 °C before demolding can occur. The advantage of this technology is that foams with low densities and high tensile strength properties can be produced, but the energy costs for heating and cooling the molds are high. This technology has lost its appeal, primarily due to high energy costs and reduced design flexibility.

The molds are usually made of aluminum, but molds of epoxy and polyester resins are also used. To achieve easy demolding, the molds must be treated with a release agent before injecting the reactive mixture. The release agents are generally highly filled water-based wax dispersions sprayed onto the mold surface. The applied release agent must match the surface properties of the PU system and the mold surface material.

The weight of material required to fill the mold is usually significantly higher than the minimum amount required to fill the mold – the just-fill weight. Overpacking can be as high as 50% to obtain a foam with a good surface definition. With increasing overpacking, the pressure in the mold increases and can reach 200 kPa. Slow reactivity and high mold pressures increase the demolding time, which may vary between 2 and 5 min.

Inserts made from metal or plastic can be placed in the mold to stiffen or reinforce the foam. Foam-in-fabric, also known as foaming-in-place, is a technology where the foam is directly molded onto a specially designed cover. The reaction mixture is then poured into a mold lined with a fabric. After demolding, the foam is precisely fitted to the cover material. This production method saves time compared to traditional cut-and-sew technology for covering foam parts.

Using more than one mold per dispensing machine can improve foam production productivity and reduce processing costs. Examples include rotary tables or carousel systems equipped with multiple molds and linked to a single fixed dispenser.

8.9.2 Slabstock foam

Slabstock foam is produced on continuous foam machines [Fig. 8.14]. The reactants are mixed and poured onto a moving conveyor belt, where the foam expands and solidifies. The continuous foam, with widths and heights of up to 2.5 and 1.5 m, is cut to lengths between 15 and 120 m. The long slabs are put into interim storage in a curing chamber to complete the reaction. Subsequently, these large buns are cut into smaller, more manageable blocks, for instance, with a length of 1 – 3 m, known as short blocks. The foams are typically produced at densities ranging from 15 to 50 kg/m^3 and can be fine- or coarse-cell. The main application areas of slabstock foam are upholstered furniture, bedding, and transportation. They are also applied in smaller-volume applications, such as kitchen sponges, air filters, and clothing.

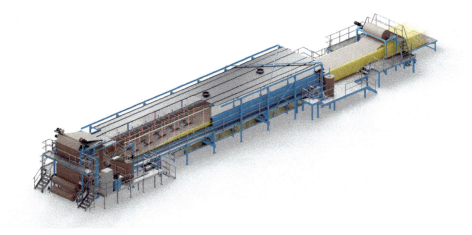

Fig. 8.14: Scheme of a continuous slabstock line (with the kind permission of Hennecke GmbH).

The foams can be polyether- or polyester-based, with the production volume of the former being significantly larger than that of the latter. The polyether slabstock foams are categorized into two types of foam: Conventional and High Resilience (HR) foam. The first is produced using slab polyol, while the second uses molded polyol. In addition to this basic differentiation, various types of slabstock foams are distinguished: Low-Density (LD), High-Load Bearing (HLB), combustion-modified ether (CME), viscoelastic (VE), and super or hypersoft foams.

TDI is usually the isocyanate of choice, but MDI is also used, for example, to produce special foam grades with higher hardness and density. Slab polyols are always combined with TDI, whereas molded polyols can be used with both isocyanates. Slab polyols require strong gelling catalysts, such as stannous octoate, in combination with amine catalysts and more strongly stabilizing surfactants. Molded polyols show intrinsically higher gelling activity and require less catalysis.

The starting materials and additives, including polyols, isocyanate, water, surfactant, catalyst, and further additives, are stored in separate tanks. The components are fed into a multi-component high or low-pressure mixing head in the preselected quantity ratio. The mixing head discharges the reaction mixture directly on the conveyor belt (the liquid lay-down process) or in a trough (the trough process). In the trough process, the reactive mixture passes over the forward edge of the trough and onto the conveyor. The conveyor and the side walls move at the same speed, forming a continuous channel lined with protective paper sheets or plastic foils. The creaming starts within seconds, the foam rise is completed within 1 – 2 min, and after 5 – 10 min, the foam bun is strong enough to be cut and handled.

As the foam begins to rise, the sidewalls exert frictional drag on the foam. This causes the foam near the sidewall to rise more slowly than in the foam core, resulting in a dome-shaped foam. Dome-shaped foams, however, produce relatively high amounts of scrap foam because the foam samples cut from these buns are mostly rectangular. To address this fundamental shortcoming, several rectangular shape (RS) production systems have been developed to produce rectangular blocks.

RS-system/Petzetakis process
The Rectangular Shape system diminishes the friction with the side walls by adding an extra protective plastic film between the sidewall paper and the foam. This film is pulled upward during the foam rise, preventing the foam from sticking to the side paper and forming a rectangular bun.

Planiblock/Flat-top process
There are several different versions of rectangular bun processes, known, for example, as Planiblock or Flat-top systems. Rectangular blocks can be produced using a fixture that mechanically compresses the top surface of the rising foam. When the foam starts creaming, an additional release paper is fed onto the surface and pressed onto it using pressure-regulating metal or wooden plates. The pressure is kept until shortly after the foam has reached its maximum height and enters the heated tunnel.

Vertifoam process

The Vertifoam process is the least common production method. The process is fundamentally different, allowing for the production of round and rectangular blocks. The foam reaction mixture is introduced at the bottom of an enclosed, vertical, upward-moving conveyor lined with paper or foil. The foam is supported by the conveyor on all sides and rises at a controlled speed dictated by the foam's expansion and cure rate. The plug flow ensures a stable and uniform flow front. When the foam leaves the conveyor, the foils are removed, and the foam is cut into cylinders or blocks. The Vertifoam line is compact and less expensive compared to the horizontal laminators. Furthermore, it can run at low output rates, and since the foams do not exhibit dense skins, less waste is generated.

8.10 Technical foams

Several foams do not fall directly into the rigid and flexible foam categories and are classified as technical foams. Commercially important technical foams are semi-rigid, viscoelastic, and packaging foam. Their physical properties and recipes are generally closer to those of flexible foam than those of rigid foam; therefore, they are discussed in this chapter.

8.10.1 Semi-rigid foams

Semi-rigid foams are predominantly used in automotive applications. The foams are open-cell but significantly harder than flexible foams, exhibiting excellent energy-absorbing properties. This makes them ideally suited for applications where high impact is required, such as safety-related applications like instrument panels, door side panels, knee bolsters, and energy-absorbing foam. Due to their open-cell nature, semi-rigid foams exhibit excellent sound insulation properties. The foam density and hardness requirements depend strongly on the application [Tab. 8.2].

Tab. 8.2: Property requirements for typical technical foams.

Application	Foam density (kg/m³)	Hardness CLD-40 (kPa)
Semi-rigid foams	100–200	25–250
Energy absorbing foams	50–100	150–1,500
Headliners	20–40	100–200

Instrument panels [Fig. 8.15] consist of a rigid insert, a soft decorative skin, and semi-rigid foam. The foam is used to bond the two substrates to form the part and to provide a soft touch. The flow path of the foam to fill the cavity formed by the insert and skin may be long and show obstacles. Therefore, foam systems that exhibit good flow and foam stability are required. An overpack of up to a factor of two may be employed to achieve complete mold fill and good adhesion. The isocyanate used is PMDI or a PMDI-based prepolymer. The polyol blend comprises water to expand the foam, molded polyols for fast reactivity and foam stability, and, optionally, crosslinkers. The crosslinkers shift the glass transition to higher temperatures and broaden the glass transition, increasing foam hardness at service temperature.

Fig. 8.15: Instrument panel (BMW) (with the kind permission of Krauss Maffei Technologies GmbH).

Energy-absorbing (EA) foams are used to minimize or completely prevent damage upon impact. The foam is part of an assembly, such as a car bumper, sandwiched between a tough outer skin, e.g., made from polypropylene, and a rigid insert. Upon collision, the impact energy is converted into work of deformation and heat. Depending on the foam formulation, EA foams can exhibit ductile viscoelastic behavior, where the foam slowly recovers after impact. However, it can also display permanent non-recoverable deformation because the cell structure is crushed. EA foams can be produced as molded foams, inserted into the casing, or, more commonly, by direct foam filling of a previously prepared casing, thereby achieving a perfect bond between the foam and the substrates.

Headliner foam belongs to the group of Thermoformable foam (TF foam). These foams are used in lightweight applications where the foam has to cover large areas. The foams are produced from formulations containing high amounts of water to achieve low densities, short-chain polyols to provide rigidity, and optionally molded polyols for softening the foam. PMDI is commonly used as the isocyanate. Large buns of TF foam are produced in a batch-block or slabstock foaming process and sliced into sheets of approximately 1 – 2 cm in thickness. A layered composite is produced in which the foam sheet is sandwiched between a reinforcing layer of, e.g., glass fleece and a decorative fabric. The sheet is then cut to size and thermoformed into the desired shape at approximately 130 °C.

8.10.2 Viscoelastic foams

PU viscoelastic foams are divided into Pneumatic Viscoelastic (PVE) and Chemical Viscoelastic (CVE) foams. These foams are produced with the same hardness-density ratio as flexible foam (CLD-40 and density of approximately 1 – 4 kPa and 35 – 80 kg/m^3, respectively) but exhibit high damping. The ball rebound is a maximum of 15%, and its recovery time after deformation is approximately 10 s [Fig. 8.16].

Fig. 8.16: Slow recovery viscoelastic foam (with the kind permission of BASF SE).

PVE foam, in its pure form, is a closed-cell foam with microporous windows. The viscoelastic effect is caused by the restricted air movement through the microporous membranes upon deformation. This type of pseudo-viscoelastic behavior is independent of temperature and is determined by the breathability of the foam.

CVE foam, in general, is open-cell, and upon deformation, it behaves like a dampened spring – its deformation and recovery are retarded. The viscoelastic effect is obtained by polymer modification whereby the glass transition temperature is shifted such that the maximum value of the loss factor is positioned between -10 and +10 °C. This means the foam is in its glass transition under environmental conditions. Consequently, the foam becomes increasingly softer and elastic with increasing temperatures and stiffer and more dampening at decreasing temperatures. The polymer of a CVE foam is crosslinked, and the foam does not exhibit permanent deformation under prolonged loading. The retardation upon deformation is known as stress relaxation and is responsible for the pressure-relieving properties of viscoelastic foam. Furthermore, it can closely conform to the shape of the human body and distribute pressure over a wider area. These properties make the foam suitable for producing mattresses, pillows, and orthopedic products. High-comfort mattresses often feature a load-bearing, high-resilience, or conventional foam base combined with a top layer of viscoelastic foam.

There are several formulation technologies for CVE and PVE foams. Often, a combination of ethylene and propylene oxide-rich polyether polyols is used. MDI is, in most cases, the preferred isocyanate. Viscoelastic topper pads and comfort layers for mattresses are produced in a slabstock process, whereas pillows and orthopedic products are molded.

8.10.3 Packaging foams

Flexible and semi-rigid polyurethane foams are used in protective packaging applications. The foams act as shock absorbers, protecting a moving or falling packaged object from damage as it encounters resistance, slows down, and finally comes to rest. Bun stock foams can be customized by impact absorption characteristics and shape, e.g., for bracing, supporting, and wrapping objects. They are often the material of choice for smaller, lighter, and more shock-sensitive products.

PU foam buns are available in a density range of $12 - 30 \text{ kg/m}^3$ and can be cut to shape for specific packaging applications. Block-shaped foams can be contoured to fit the article into the cut slots, and then the top and bottom are backed with foam sheets. Soft foam sheets cut from rolls can be wrapped around items for protection. Convoluted or egg box foam is a packaging solution comprising PU foam plates with regular and repetitive peaks and troughs. They are generally used in pairs, with one layer at the top and the other at the bottom. The system encloses items and maintains their position during handling and transportation.

Foam-in-place packaging is an in-situ packaging technology where polyurethane (PU) foam resin is injected into a plastic bag, and the expanding foam surrounds the object, filling the outer container. The foam formulation can be adjusted to obtain the desired hardness-density properties. Foam densities as low as 5 kg/m^3 can be achieved.

References

[1]   E.G. Rightor, S.G. Urquhart, A.P. Hitchcock, H. Ade, A.P. Smith, G.E. Mitchell, R.D. Priester, A. Aneja, G. Appel, G. Wilkes, W.E. Lidy, Identification and Quantitation of Urea Precipitates in Flexible Polyurethane Foam Formulations by X-ray Spectromicroscopy, Macromolecules, 2002, 35, 5873–5882.

[2]   W. Li, A.J. Ryan, I.K. Meier, Morphology Development via Reaction-Induced Phase Separation in Flexible Polyurethane Foam, Macromolecules, 2002, 35, 5034–5042.

[3]   M.J. Elwell, A.J. Ryan, H.J.M. Grünbauer, H.C. Van Lieshout, In-Situ Studies of Structure Development during the Reactive Processing of Model Flexible Polyurethane Foam Systems Using FT-IR Spectroscopy, Synchrotron SAXS, and Rheology, Macromolecules, 1996, 29, 8, 2960–2968.

[4]   J.H. Saunders, The formation of urethane foams, Rubber Chem. Technol., 33, 1293–1322 (1960).

[5]   L.D. Artavia, C.W. Macosko in Low density cellular plastics, N.C. Hilyard, A. Cunningham, Eds, Chapman and Hall, London (1994) ISBN 0 412 58410 7, Chapter 2.

[6]   R.A. Neff, C.W. Macosko, Simultaneous measurement of viscoelastic changes and cell opening during processing of flexible polyurethane foam, Rheol. Acta, 35, 656–666 (1996).

[7]   R. Herrington, K. Hock in Flexible Polyurethane Foams, R. Herrington and K. Hock, Eds, Dow Chemical, Midland, 1997, Chapter 3.

[8]   W.E.A. Davies, The theory of composite dielectrics, J. Phys. D: Appl. Phys., 4 (1971) 318–328.

[9]   A.J. Ryan, J.L. Stanford, R.H. Still, Thermal, mechanical and fracture properties of reaction injected moulded poly(urethane-urea)s, Polymer, 32, 1426–1439 (1991).

[10] L.J. Gibson, M.F. Asby, Cellular Solids, 2nd Ed., 1997, Cambridge University Press, ISBN 0521-49560-1, Chapter 5.

[11] N.C. Hilyard, A. Cunningham in Low Density Cellular Plastics, N.C. Hillyard and A. Cunningham, Eds, Chapman and Hall, 1994, ISBN 0-412-58410-7, Chapters 1 and 8.

[12] G. Verhelst, A. Parfondry, N. Duggan in The Polyurethanes Book, D. Randall and S. Lee, Eds, Wiley, 2002, ISBN 0-470-85041-8, Chapter 11–14.

9 Elastomers

Elastomers are polymers that exhibit rubber-like elasticity at ambient temperatures; after deformation, they regain their original shape. The deformation properties are dependent on the polymer hardness. Soft elastomers typically exhibit high elasticity and extensibility, whereas hard elastomers display high toughness.

PU elastomers show a phase-separated structure of hard and soft domains. The soft domains provide elasticity, whereas the hard domains act as physical crosslinks, providing mechanical strength. PU elastomers can be thermoplastic; however, most PU elastomers are thermosets. Thermoplastic polyurethanes (TPU) only have physical crosslinks that melt at high temperatures, allowing melt processing. Thermoset PU elastomers are produced in their final form and feature a combination of physical and chemical crosslinks. PU elastomers can be foamed or compact, with densities ranging from 150 to 1,100 kg/m^3.

With the large number of available starting materials, polyurethane elastomers can cover a wide range of properties and are used for numerous applications [Fig. 9.1]. The most important PU elastomer applications can be grouped into the following categories: cast, microcellular, RIM and spray elastomers, artificial leather, elastomeric fibers, and TPU. PU elastomers are produced in two-component molding processes, whereas TPU is melt-processed.

Fig. 9.1: Some applications of PU elastomers: Shoes, cast elastomers, and jounce bumpers (with the kind permission of BASF SE (left) and Hennecke GmbH (right)).

Cast elastomers are compact polymers that can be produced at various hardness levels and are used in manufacturing a wide range of engineering parts, such as rollers for roller skates and wheels for specialty vehicles, such as lawnmowers and forklift trucks.

Microcellular elastomers are molded high-density elastomer foams with densities ranging from 200 to 800 kg/m^3. The cells are much finer than those of low-density foams and are usually closed. The major application for microcellular foams is in shoe soling. Furthermore, they are used in technical foam applications such as steering wheels, auxiliary springs for the automotive industry, and sheets for vibration isolation.

RIM (Reaction Injection Molding) elastomers exhibit densities above 900 kg/m^3 and are thus virtually solid. RIM parts are often reinforced using filler materials, giving highly dimensionally stable articles for the automotive industry, such as body panels, bumpers, spare wheel covers, and truck floors.

Spray elastomers are applied as coatings. The primary application areas for spray elastomers are in the construction industry as protective coatings. Typical examples are flooring, secondary containment, and anti-corrosion coatings for metal containers.

Artificial leather is produced using a coagulation process, which yields breathable textiles used, for example, in rain and outdoor clothing. PU fibers are used in underwear and swimming suits where high elasticity of the woven textile is required.

TPUs are supplied to manufacturers as fully reacted polymers produced in TPU production plants. The TPU granules are processed using standard injection molding equipment. Due to their high flexibility, abrasion and oil resistance, TPUs are used in applications such as cables, hoses, membranes, and various engineering parts, including mining sieves and belts. Dissolved in organic solvents, TPUs can be used as adhesives or coatings.

9.1 Elastomer starting materials and formulations

Polyurethane elastomers are produced from diisocyanates, long-chain diols, and chain extenders.

9.1.1 Diisocyanates

Predominantly aromatic diisocyanates are used for the production of elastomers [Chapter 2.1]. The standard product is 4,4'-MDI, followed by TDI (2,4- and 2,6-TDI at an 80:20 ratio), 1,5-naphthalene diisocyanate (NDI), and 3,3'-dimethyl-4,4'-biphenyl diisocyanate (TODI). Aliphatic isocyanates are also applied, albeit in much smaller volumes. Of the aliphatic isocyanates, HDI is used the most, followed by IPDI, H$_{12}$MDI, and TMXDI.

TDI is primarily used for soft elastomers, while NDI and TODI are suitable for high-performance elastomers. The rigid aromatic diisocyanates form strong hard domains, providing elastomers with excellent physical properties. The hard domains from the more flexible and bulkier aliphatic diisocyanates are less organized, resulting in elastomers with weaker mechanical properties. An exception is HDI. The even-numbered straight aliphatic diisocyanate allows for easy stacking, and the elastomers exhibit strain-induced hardening [1]. Elastomers based on aliphatic isocyanates are primarily used in applications that require lightfastness, UV stability, and transparency.

9.1.2 Polyols

The most commonly used polyols for elastomers are polyether and polyester polyols, with a functionality of 2 – 3 and equivalent masses ranging from 500 to 2,000 g/mol. The glass transition temperature of the neat polyols typically ranges from -70 to -30 °C. The polyether polyols comprise polypropylene glycol (PPG), reactive or molded polyols (propylene oxide-based cores with ethylene oxide-based caps), and polytetramethylene ether glycol (PTMEG).

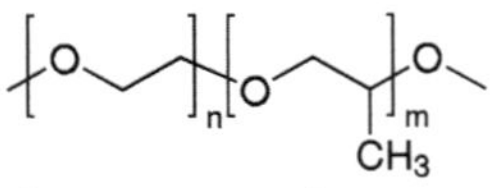

EO/PO polyether

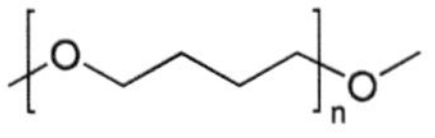

Polytetrahydrofurane

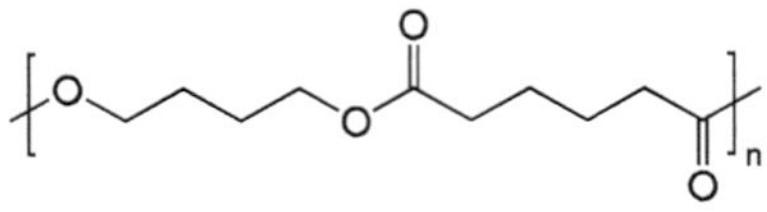

Poly(1,4-butylene adipate)

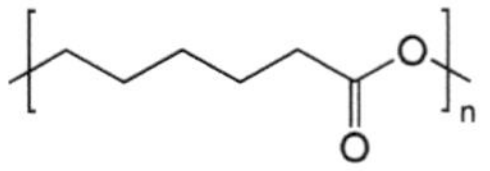

Polycaprolactone

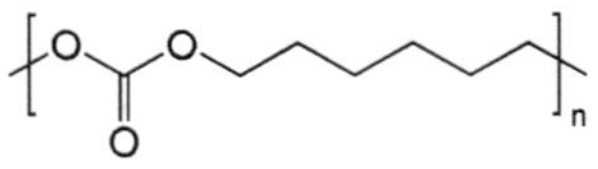

Poly(hexamethylene carbonate)

The polyester polyols used for manufacturing PU elastomers are based on adipic acid and short-chain diols with primary hydroxyl groups, such as ethylene glycol, 1,4-butanediol, 1,6-hexanediol, and diethylene glycol. Depending on the diol employed and the molar mass of the polyol, adipate polyester polyols can be crystalline. Their melting points increase with the molar mass of both the diols used and the polyol produced. Liquid noncrystalline polyesters, however, are preferred for PU elastomer applications. This can be achieved by selecting the appropriate glycol, using mixtures of glycols, or reducing the molar mass of the polyol.

PTMEG, polycaprolactone, and polycarbonate diols based on 1,6-hexanediol are used to produce high-performance PU elastomers.

9.1.3 Chain extenders

The chain extenders used in PU elastomers are low-molecular-mass diols with primary hydroxyl groups. The most commonly used chain extender is 1,4-butanediol. Furthermore, ethylene glycol, 1,3-propanediol, diethylene glycol, and hydroquinone-bis(2-hydroxyethyl) ether are used. Diamine chain extenders are utilized in certain specialized elastomer applications. For processing reasons, only aromatic diamines with reduced reactivity are suitable. Reduced reactivity is achieved by applying steric hindrance on the amine group or using electron-withdrawing substituents. The most commonly applied aminic chain extender is DETDA, a mixture of 2,4- and 2,6-diethyltoluene diamine.

9.1.4 Basic recipe

Tab. 9.1 presents a basic recipe for PU elastomers, with the polyol amount set at 100 parts by weight. The chain extender amount can vary widely. Its amount determines the hard block content (HBC), which, in turn, determines the hardness of the PU elastomer.

Tab. 9.1: Basic PU elastomer recipe.

Component	Amount (p.b.w.)
Polyol	100
Chain extender	2–20
Water	0–1
Catalyst	0–0.5
Diisocyanate	30–80
Index	100–120

The addition of water is optional. Formulations for solid elastomers are typically devoid of water, whereas recipes for shoe soling applications may contain up to 1 part of water per 100 parts by weight of polyol. The most important reaction for manufacturing two-component PU elastomers is the urethane reaction. Therefore, urethane-selective catalysts such as triethylenediamine (TEDA) or metal salts, such as tin carboxylates, are used.

The index is generally close to 100. Higher indices up to 120 can be applied when allophanate crosslinking is desired. The above model formulation would require 30 – 80 parts by weight of isocyanate when using 4,4'-MDI.

The HBC, expressed as weight percent, can be calculated from the weight amounts of isocyanate (w_{iso}), chain extender (w_{CE}), and polyol (w_{polyol}) as shown on next page. The calculation does not account for the presence of water in the formulation.

$$HBC = \frac{100 \cdot (w_{iso} + w_{CE})}{(w_{polyol} + w_{iso} + w_{CE})}$$

9.2 Chain topology

The polymerization of PU elastomers in industry is carried out in bulk using liquid reaction components. The chain topology, which is the structure of alternating sequences of diisocyanate-co-chain extender and polyol segments, depends on the elastomer recipe and the polymerization conditions. The discussion on chain topology takes TPU as an example but can be extended to crosslinked elastomers.

Let us consider a basic TPU recipe with the following composition: one equivalent polyol (M = 2,000 g/mol), two equivalents of chain extender (M = 100 g/mol), and, to arrive at an index of 100, three equivalents of diisocyanate (4,4'-MDI; MW = 250 g/mol). The HBC of this polymer is 32 wt.%, and the molar ratio of polyol, chain extender, and isocyanate amounts to 1:2:3 [Tab. 9.2].

Tab. 9.2: Recipe of the basic TPU (index = 100).

	M (g/mol)	EM (g/mol)	Amount (p.b.w.)	Equivalents	Molar ratio
A-component					
Polyol	2,000	1,000	100	0.1	1
Chain extender	100	50	10	0.2	2
B-component					
4,4'-MDI	250	125	37.5	0.3	3

Reacting the three components, a block-copolymer of the [A-B]$_n$ type is formed, with a hard block (B) consisting, on average, of three isocyanates and two chain extender molecules connected through polyol soft blocks (A) [Fig. 9.2].

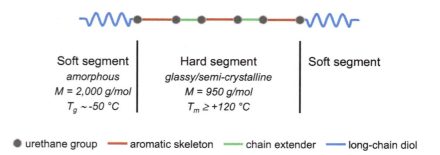

Fig. 9.2: Idealized topology of a segmented polyurethane chain with HBC = 32 wt.%.

The average molar mass of the soft segment in our example is 2,000 g/mol, and that of the hard segment amounts to 950 g/mol arithmetically. Thus, the average repeating unit has an average molar mass of 2,950 g/mol. TPUs typically have a molar mass of approximately 100,000 g/mol. Hence, the polymer chains in the present example contain approximately 30 alternating hard and soft segments [Fig. 9.3].

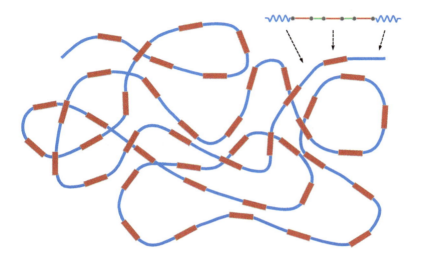

Fig. 9.3: "Average" topology of a TPU block copolymer with an HBC of 32 wt.% and a molar mass of 100,000 g/mol.

The hard blocks, however, exhibit a broad molar mass distribution, consisting of both short and long hard block sequences that are statistically distributed along the polymer chain. Under ideal polymerization conditions, the hard blocks may exhibit a "most probable" mass distribution, a probability distribution according to the Schulz-Flory distribution. Let us consider the hard block structures in a PU elastomer produced from, for example, PTMEG, 4,4'-MDI, and 1,4-butanediol. Possible buildup sequences of hard chain segments and their likelihood of being involved in forming hard domains are presented in Tab. 9.3. A hard block with chain extender number (0), i.e., an MDI molecule that has reacted on both sides with a polyol, will remain in the soft phase. Depending on the temperature conditions, the hard block consisting of two MDI molecules and one chain extender (1) might be dissolved in the soft phase, but it can also be part of a hard domain, whereas the hard blocks (2) and higher will most likely be involved in building hard domains [2].

The hard and soft blocks are thermodynamically incompatible, resulting in phase separation and the formation of polyol and hard block-rich domains. The phase separation process is polymerization-induced. Once the hard blocks exceed a critical mass, they become incompatible with the surrounding liquid and phase separate. The conversion at which phase separation occurs depends on the starting components, reaction rate, and temperature.

Tab. 9.3: Possible buildup sequences of the hard blocks in TPU. M, C, and S represent 4,4'-MDI, chain extender, and polyol, respectively.

Hard block structure	Number of chain extenders	Involvement in hard domain
S-M-S	(0)	No
S-M-C-M-S	(1)	Maybe
S-M-C-M-C-M-S	(2)	Yes
S-M-C-M-C-M-C-M-S	(3)	Yes
S-M-[C-M-]$_n$-S	(n $\geq$ 4)	Yes

9.3 Polymer morphology

The hard and soft segments in PU are thermodynamically incompatible, resulting in phase separation and the formation of polyol and hard block-rich domains. The soft phase is generally the continuous phase in which the hard domains are dispersed. The intramolecular forces between the hard blocks ensure the formation of stable hard domains that melt at high temperatures. The hard phases fulfill a dual role, acting as both physical crosslinks and fillers, which provide recovery upon extension and increase polymer hardness.

9.3.1 Chain statistics and structural parameters

The stacking of the hard blocks in the hard domains shows imperfections related to chain statistics, structural parameters, and processing.

The arrangement of polymer chains in a TPU can be envisioned as a "spaghetti-like" entangled polymer mass in which a single chain can be involved in various hard domains. The entangled structure and cooperative effect of hard blocks within a particular chain being part of several different hard domains provide polymer strength. The ordering of the hard blocks in the hard domains is imperfect, given the broad length distribution of the hard blocks. Fig. 9.4 schematically illustrates the mode of hard block association within the hard domains. The portrayed hard domain shows the effect of heterodispersity and dislocation of the hard blocks. A distinction can be made between isocyanate-chain extender repeating units that show lateral ordering and build the "core" of the hard domain and units that end up in the "interphase". Only the former provides hardness and strength to the polymer. Thus, for an elastomer with a given HBC, a well-ordered hard domain will show a higher hardness than a poorly ordered structure containing more interphase material.

The extent to which phase separation occurs depends on the hard block content, the polarity and compatibility differences between the hard and soft blocks, and the ability of the hard blocks to stack and establish intermolecular lateral interactions. The stackability

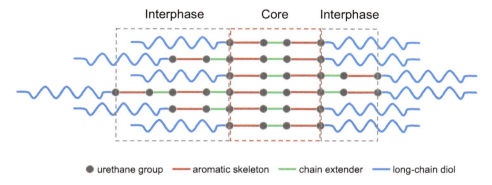

Fig. 9.4: Schematic representation of hard segment stacking, illustrating the effect of heterodispersity and dislocation of the hard blocks on core and interphase formation.

of the hard blocks depends on the structure of the isocyanate and chain extender from which they are made. A strong lateral interaction between the hard blocks gives stable hard domains with high melting temperatures. An increase in HBC increases the size and, concomitantly, the stability of the hard domains. Phase mixing can occur when hard and soft blocks show considerable compatibility, whereas poor compatibility increases the extent of phase separation. The segregation between the phases is enhanced with increasing differences between the glass and melting temperatures of the soft and hard phases [3]. This implies that hard blocks, which can develop strong lateral interactions and high levels of order in combination with apolar, low T_g polyols, yield elastomers exhibiting high levels of phase separation.

9.3.2 Hard segment interactions

The interactions between the hard blocks in the hard domains can be of different origins. The primary interaction between hard segments involves the formation of hydrogen bonds between the carbonyl oxygen and the urethane N-H. The ability to form hydrogen bonds depends on the geometry of the hard blocks and the HBC. The former determines how well the hard blocks can stack, whereas the latter governs the length of the hard blocks.

The inter-hard block interactions were studied using molecular modeling on methanol-capped 4,4′-MDI model hard segments [4]. Four types of interactions were relevant for establishing cohesive interactions between the hard blocks [Fig. 9.5].

The strongest interaction is the N-H⋯O=C hydrogen bond between two urethane groups [Fig. 9.5a]. Furthermore, it was shown that the free electron pair at the single-bonded urethane oxygen can also serve as a hydrogen bond acceptor [Fig. 9.5b]. The dipole-dipole interactions between the urethane carbonyl groups exhibit the largest

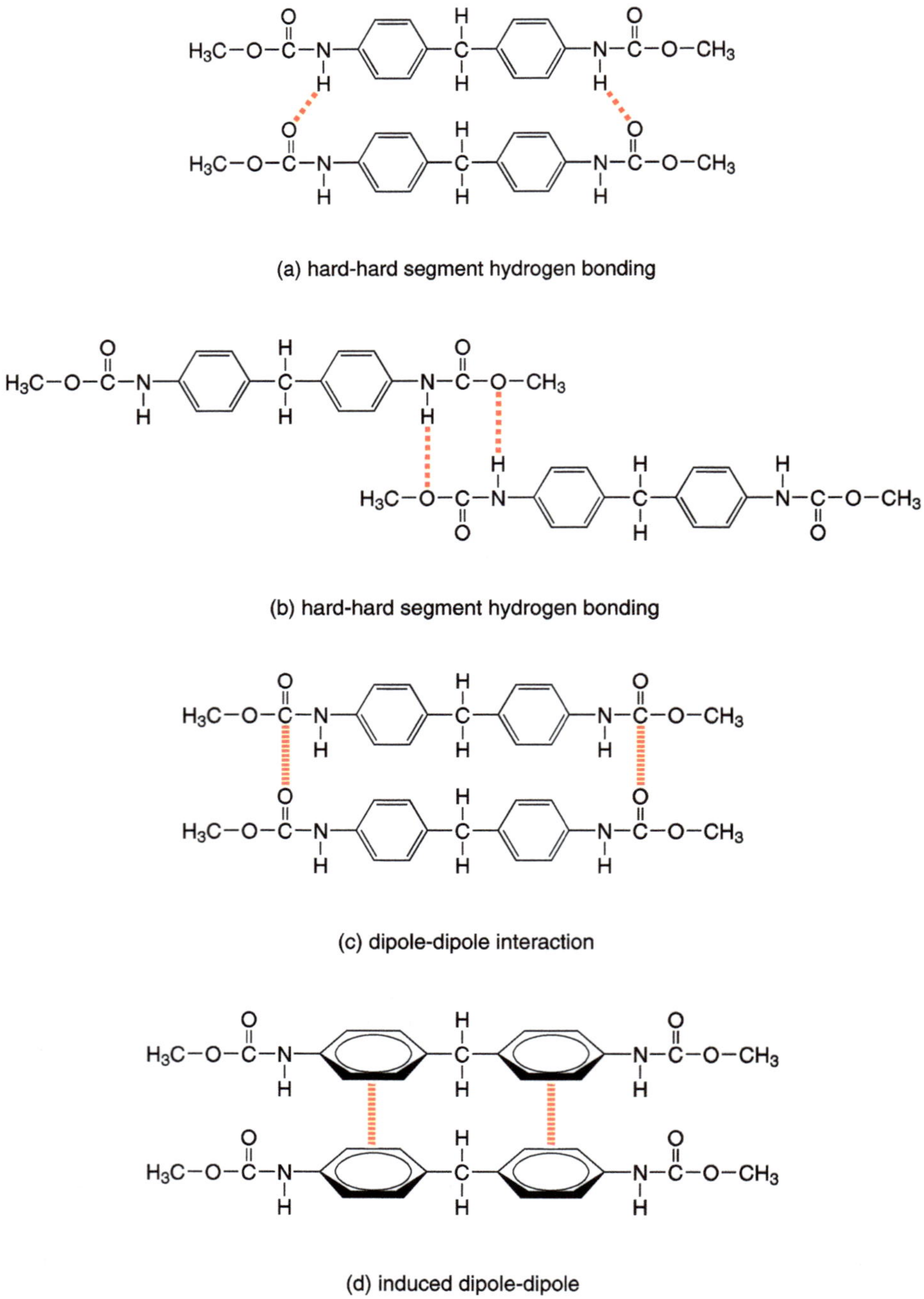

198 —— 9 Elastomers

(a) hard-hard segment hydrogen bonding

(b) hard-hard segment hydrogen bonding

(c) dipole-dipole interaction

(d) induced dipole-dipole

Fig. 9.5: Possible interactions between methanol-capped MDI model hard segments, adapted from [4] (with the kind permission of Springer Nature).

dipole moment [Fig. 9.5c]. The induced dipole-dipole interactions between the aromatic rings possess a small dipole moment [Fig. 9.5d].

A strong hydrogen-bonded network of urethane groups in the hard domains is fundamental to the strength properties of a PU elastomer. A qualitative indication of the degree of hydrogen bonding can be obtained from IR spectroscopy. The IR absorptions are highly sensitive to hydrogen bonding in the carbonyl region, as evidenced by peak shifts. The unbound carbonyl bonds of urethane and urea exhibit strong absorptions at 1,730 and 1,710 cm^{-1}, respectively, which shift to lower frequencies upon the formation of hydrogen bonds. This effect was studied in a series of model PU elastomers based on polypropylene glycol (M = 2,000 g/mol), 1,4-butanediol, and 4,4′-MDI at different polyol and chain extender ratios, with an index of 100. The HBC in this series of elastomers was increased from 11 to 21, 23, and ultimately to 25 wt.%. The amid-1 region of the spectra shown in Fig. 9.6 was deconvoluted into "free" and "hydrogen-bonded" urethane at 1,730 and 1,710 cm^{-1}, respectively.

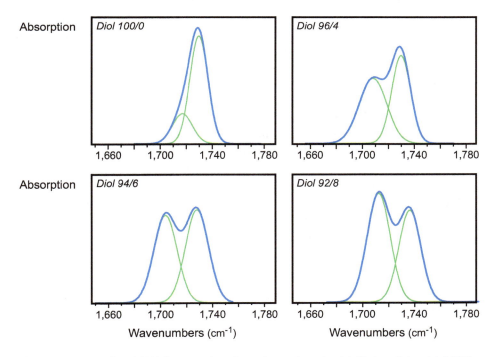

Fig. 9.6: IR spectra of model PU elastomers based on polypropylene glycol, 1,4-butanediol, and 4,4′-MDI at various polyol-to-chain extender ratios.

The elastomer without a chain extender exhibits hard blocks of type (0) [Tab. 9.3]. These blocks are too short to induce phase separation and remain dissolved in the soft phase. Despite being dissolved, some urethane-urethane hydrogen bonding occurs

[Fig. 9.6 upper left]. The length of the hard blocks, and concomitantly the size of the hard domains, increases with increasing amounts of chain extender. As a result, hydrogen bonding becomes stronger. The increased hydrogen bonding and hard domain strength were evidenced by the evolving absorption at 1,710 cm^{-1} in IR and the increase in tensile strength in Fig. 9.18 (page 218), respectively.

9.3.3 Polyol requirements for phase separation

The polarity of the polyol and its molecular mass determine the soft phase structure. Low molar mass and high polarity favor phase mixing, while high molar mass and low polarity promote phase segregation.

It is generally accepted that the minimum molar mass for the commonly used polyether and polyester polyols in MDI-based elastomers to induce phase separation is approximately 600 g/mol. Only at molar masses higher than 600 g/mol is the thermodynamic drive for phase separation sufficient to enable phase separation. The extent of phase separation increases with the increasing molar mass of the polyol. The soft phases get thicker, and the soft phase T_g decreases, improving elastomer resilience and low-temperature flexibility. At the same time, the hard blocks become longer and the hard domains thicker and stronger. As a result, both the polymer modulus at room temperature and the melting temperature increase. The reduced T_g and increased T_m extend the modulus plateau region.

A simple calculation illustrates the formation of longer hard blocks when the molar mass of the polyol is increased. Let us reconsider the basic TPU formulation in Tab. 9.2; page 194. When reducing the molar mass of the polyol from 2,000 to 1,000 g/mol, the chain extender amount must be reduced to approximately 6 p.b.w. to maintain the HBC at 32 wt.%. Then, the "average" hard segment consists of 1.6 mol of 4,4'-MDI and 0.6 mol of chain extender, with a molar mass of 460 g/mol. The calculated molar mass is approximately half that of the TPU, based on the polyol with a molar mass of 2,000 g/mol. The mass of the hard blocks, calculated for the two polyols, is depicted in Fig. 9.7.

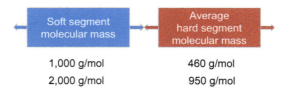

Fig. 9.7: Average molecular mass of hard blocks calculated for two polyols with different molar masses (1,000 and 2,000 g/mol) at an HBC of 32 wt.%.

The relatively nonpolar polyol soft blocks are thermodynamically incompatible with the more polar urethane-containing hard blocks. This incompatibility fundamentally drives the phase separation process. When the polarity difference between the two blocks decreases, the phases will become more phase-mixed.

Polyether and polyester polyols differ in polarity. The free electron pair of the ether oxygen acts as a weak hydrogen bond acceptor, whereas the ester group forms hydrogen bonds via the carbonyl and the ester alkoxide. The ester bond is, therefore, a more polar and stronger hydrogen bond acceptor than the ether bond. Consequently, polyether polyols, such as those based on polypropylene glycol and PTMEG, are less compatible with the hard blocks than common adipate-based polyester polyols. Therefore, polyester elastomers at similar molar masses of the polyol show more phase mixing. This results in a broader soft-phase transition and an increased T_g.

The polarity of the polyols, and hence their tendency to phase mix, can be reduced by increasing the number of carbon atoms between the polar groups. The molecular mass distribution of the polyol also affects phase separation, as a broader distribution enhances phase mixing. Polyester polyols generally show a broader polydispersity index than polyether polyols, increasing their tendency to phase mix.

Polyethylene glycol is a special case. The two adjacent ether oxygens, connected by a two-carbon bridge, can form strong bidentate-type hydrogen bonds [Fig. 9.8] to the NH group of a urethane. Because of this strong interaction, the soft and hard blocks are miscible, and phase separation does not occur. Phase-mixed PUs are soft and weak, which is one of the reasons why PEG polyols are not used in PU elastomer technology. The ether bidentate interaction does not occur in propylene glycol-based PUs. The interaction between urethane and the ether groups in polypropylene glycol, which also has two carbon atoms between the ether oxygen atoms, is much weaker due to the steric hindrance by the lateral methyl group. Ethylene oxide-capped polypropylene glycols, however, are utilized in elastomer technology. The ethylene oxide cap offers processing benefits, including increased reactivity (due to the primary hydroxyl end group) and compatibility with other starting materials in the formulation.

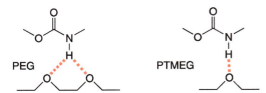

Fig. 9.8: The interaction between polyether oxygen atoms and urethane, illustrating PEG and the C_4 polyether PTMEG.

The hard phase governs the mechanical properties of PU elastomers; however, the soft and interphase also contribute. Polyester polyol systems show improved mechanical properties compared to polyethers for two reasons. First, the ester-ester interaction in the soft phase is stronger than the ether-ether interaction, and consequently, the polyester soft phase exhibits a higher cohesive energy and mechanical strength.

Second, polyester-based elastomers show a higher extent of phase mixing. This results in a higher and broader glass transition, with the loss factor at room temperature being increased. The strength and toughness of the polymer are increased, albeit at the expense of reduced resilience.

Tab. 9.4 summarizes the influence of polyol structure on the properties of PU elastomers. Generally, polyester systems exhibit superior mechanical properties, including tensile strength, tear resistance, and abrasion resistance, as well as enhanced resistance to non-polar solvents such as petrol and gasoline. However, the main drawback of polyesters is their sensitivity to hydrolysis, which may limit their applicability in humid environments. Polyether systems show superior hydrolysis resistance and low-temperature flexibility. The mechanical properties of PTMEG systems supersede those based on propylene glycol and molded polyols. However, polypropylene glycol-based polyols dominate the elastomers market because they are less expensive and generally provide sufficient properties for most elastomer applications.

Tab. 9.4: Influence of polyol structure on the properties of PU elastomers.

Property	Polyester (poly adipate)	Polyether (PO and PO/EO)	Polyether (PTMEG)
Tensile strength	excellent	fair	good
Tear resistance	excellent	fair	good
Rebound resilience	good	good	excellent
Abrasion resistance	excellent	fair	good
Hydrolysis	fair	excellent	excellent
Oil resistance	good	fair	fair
Cost	medium to high	low	medium

9.3.4 The geometrical structure of the hard block

The ability of the hard blocks to stack and form stable hard domains is largely determined by the structure of the diisocyanate and chain extender. The stacking of the hard blocks depends on the symmetry and planarity of the isocyanate, as well as the number of carbon atoms between the two hydroxyl groups of the chain extender.

The ability of the hard segments to stack depends on the symmetry and planarity of the diurethane [5]. Planar and symmetrical diurethanes can stack tightly, allowing them to develop strong intermolecular interactions. Fig. 9.9 illustrates the calculated molecular structures of the dimethyl urethanes derived from five aromatic diisocyanates: NDI, TODI, 4,4'-MDI, and 2,4- and 2,6-TDI. The first three isocyanates contain two aromatic rings, whereas TDI only possesses one. The corresponding urethanes are abbreviated 1,5-NDU, TODU, 4,4'-MDU, and 2,4- and 2,6-TDU, respectively. The geometry of the diurethanes was calculated using MNDO [6].

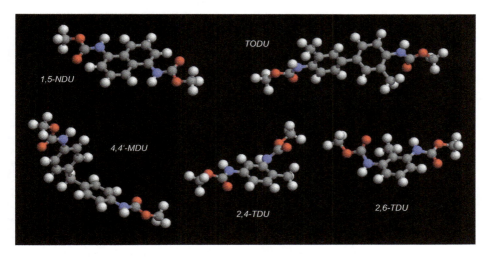

Fig. 9.9: Calculated molecular structures of the dimethyl urethanes of 1,5-NDI, TODI, 4,4'-MDI, 2,4-TDI, and 2,6-TDI.

The 1,5-NDU is planar and symmetric about the normal axis of the naphthalene scaffold and has a perfect geometry for stacking. The two aromatic rings in TODI are directly connected through a carbon-carbon bond. TODU is symmetric about an axis through the carbon-carbon bond between the two aromatic rings. However, the two aromatic rings are twisted at an angle of about 57°, according to our calculations. The proximity of the two hydrogens in the ortho position forces the molecule to twist. The tilting of the two aromatic rings will reduce the ability to stack. The 4,4'-MDU is symmetrical about an axis through the central methylene bridge but not planar, as the tetrahedral angle at the methylene group causes the molecule to bend and twist the two aromatic rings. Its kinked and twisted structure significantly reduces its ability to stack. The diurethanes of the two TDI isomers are both planar. The 2,6-TDU unit is symmetric about the central C1-C4 axis, whereas 2,4-TDU is asymmetric. In both isomers, the methyl group pushes the urethane groups out of the plane, thus impeding stacking. Perhaps even more important, TDI is generally applied as an 80:20 mixture of 2,4- and 2,6-TDI; the mixture of isomers strongly reduces the stackability of the diurethanes.

The above aromatic isocyanates, in combination with a specific polyol-chain extender combination, will yield elastomers with different hard phase structures. The ability of the hard chain segments to stack and develop strong intermolecular interactions, and concomitantly the hard phase melting temperature (T_m), would follow the descending order: NDI > TODI > 4,4'-MDI > TDI. Furthermore, hard blocks that show strong lateral interaction and high levels of order also give elastomers with high levels of phase separation. Consequently, the glass transition becomes sharper, the T_g shifts to lower temperatures, and the resilience increases. The plateau modulus increases with the strength of the hard domains and follows the above sequence. The results are summarized in Tab. 9.5.

Tab. 9.5: Development of T_g, T_m, modulus, and resilience for a series of elastomers based on a standard elastomer formulation using NDI, TODI, MDI, and TDI.

Isocyanate	T_g	T_m	Modulus	Resilience
NDI				
TODI	Increase ↓	Increase ↑	Increase ↑	Increase ↑
MDI				
TDI				

For practical purposes, elastomers have to be compared at equal hardnesses. The hardness of the softer elastomers with less perfect hard domains can be increased by increasing the HBC. The increase in HBC makes the hard domains larger and stronger; however, due to inferior stacking, the hard domains also become more diffuse, and the amount of interphase material increases. Consequently, the glass transition of the soft phase broadens and shifts to higher temperatures, resulting in a decrease in resilience. In other words, comparing a series of elastomers with different isocyanates at equal hardnesses amplifies the T_g increase and the reduction in resilience.

Numerous investigations into the effect of aliphatic glycol chain extenders on the properties of PU elastomers are available; however, no clear conclusion can be drawn from the results. One of the challenges in evaluating the effect of chain extenders is that elastomers exhibit several mechanical properties relevant to their applications. A particular chain extender may excel in one property and struggle in another, making a clear overall ranking difficult. Furthermore, the properties are system-dependent; the structure of the polyol and isocyanate, as well as the process conditions, may also affect the outcome. Despite these considerations, it can be said that glycols with an even number of carbon atoms give superior properties. In most systems, 1,4-butanediol gives the best overall properties balance [3].

The superior performance of even-numbered glycol chain extenders over odd-numbered ones is illustrated in Fig. 9.10, which compares the hydrogen bonding interaction between hard chain segments based on an even (C4 – 1,4-butanediol) and an odd-numbered (C5 – 1,5-pentanediol) glycol. The planar zig-zag conformation of the alkyl unit is the most likely possibility, leading to an acceptable chain packing. Starting from one urethane group, the orientation of the second urethane group flips over by 180° for every carbon atom added to the glycol chain. The orientation of the two urethane groups in the butanediol segment is bidirectional; the direction of the hydrogen bonding alternates back and forth, resulting in unstrained hydrogen bonds formed between neighboring hard blocks. This is different for 1,5-pentanediol. The urethane groups are oriented in the same direction, resulting in a one-directional hydrogen bonding interaction. The hydrogen bonds experience considerable strain and must adopt higher energy conformations, thus affecting the stability of the hard domain. Therefore, the highest crystallinity is achieved with even-numbered chain extenders, as the corresponding hard blocks can crystallize more easily in the lower-energy conformation and exhibit higher melting points.

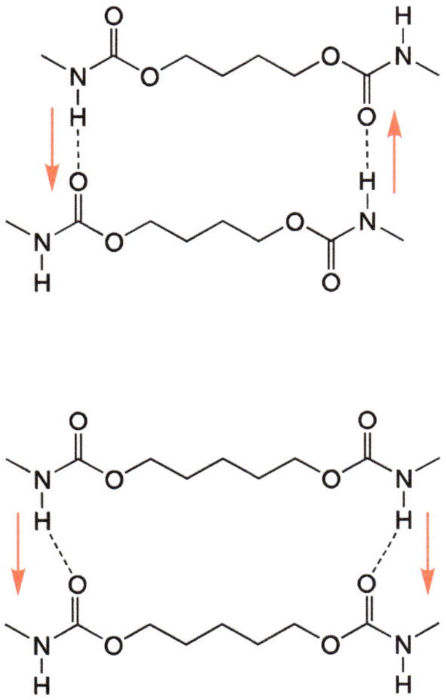

Fig. 9.10: The inter-chain interaction of hard segments with even- and odd-numbered glycol chain extenders, adapted from [7]. The direction of the hydrogen bonds is indicated by red arrows.

The odd-even effect was elegantly demonstrated in a series of [n]-polyurethanes produced from α,ω-isocyanato alcohols with different numbers of carbon atoms between the isocyanate and alcohol group [8]. The highly crystalline polymers displayed a strong odd-even effect in their melting points [Fig. 9.11]. The melting temperatures of the polymers derived from even-numbered carbon monomers were systematically higher than those derived from odd-numbered monomers. The melting points in both series of even and odd-numbered chain extenders decreased with increasing number of carbon atoms. The decline in melting point resulted from a reduced concentration of urethane linkages per unit volume, thus lowering the crystal cohesive energy. A random copolymer, based on equal molar amounts of 5-isocyanatopentanol and 6-isocyanatohexanol, however, was completely amorphous. This shows that the formation of crystalline structures is only possible for regular polymer structures.

The MDI-based hard segment structure was studied in detail. An X-ray study of 1,4-butanediol / 4,4'-MDI model compounds revealed that the MDI unit is V-shaped and can be stacked so that one urethane group can form a hydrogen bond within the same plane [Fig. 9.12]. The second urethane group of the MDI unit cannot form hydrogen bonds within the same plane but can with the groups from neighboring planes, stabilizing the structure in two directions. The urethane groups are almost planar,

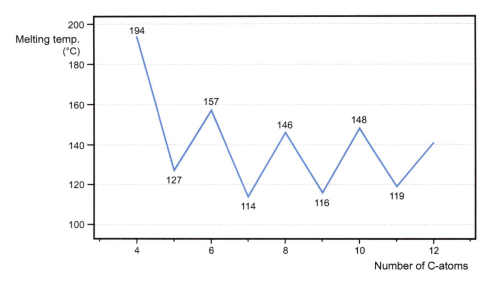

Fig. 9.11: Melting point of [n]-polyurethanes versus the number of carbon atoms between the hydroxyl and isocyanate groups, adapted from [8] (with the kind permission of Wiley and Sons).

and the carbon chain of the chain extender between the urethanes shows a planar zig-zag conformation. The phenyl groups lie in nearly perpendicular planes, and the angle of the methylene bridge between the two phenyl rings is bent out somewhat, amounting to 115°. Hydrogen bond lengths between the urethane groups were 0.289 nm perpendicular and 0.296 nm in the plane [9].

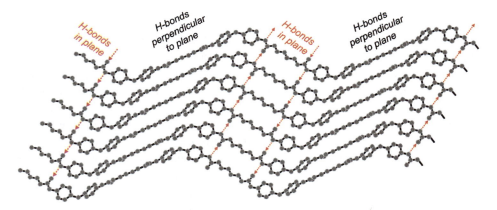

Fig. 9.12: Crystal structure of 4,4′-MDI/1,4-butanediol hard domains. The directions of the hydrogen bonds are indicated by arrows, adopted from [9] (with the kind permission of Elsevier Science Technology Journals).

With the use of diamines as chain extenders or water, the hard segment contains urea groups [10]. Urea bonds are inherently more thermally stable than urethane bonds, thereby providing elastomers with enhanced thermal properties. Because the urea group is more polar and can form two hydrogen bonds (one per N-H), the urea hard blocks form stronger hard domains, exhibiting increased levels of phase separation. Therefore, in general, urea-extended elastomers yield superior mechanical and thermal properties. The melting point of urea hard domains often exceeds the polymer decomposition temperature, and as a result, polyurea elastomers cannot be melted.

9.3.5 Physical and chemical crosslinking

Elastomers are defined as materials displaying high and reversible deformations. To achieve recovery upon deformation, they must be crosslinked. The crosslinks in PU elastomers can be of a chemical or physical origin.

TPUs are segmented, high-molar-mass polymers that exhibit an entanglement network. The hard domains, formed by the segregation of hard chain segments, provide physical crosslinks. The entanglement network and physical crosslinks ensure the excellent elastic properties of TPUs in the temperature range of 0 – 100 °C. The hard domains, however, melt at high temperatures, rendering the mechanical behavior of TPU highly temperature-dependent; close to their melting point, they deform easily at small strains.

The standard theory of rubber elasticity, which is used to analyze the elastic properties of polymer networks, may not be directly applicable to PU elastomers because the hard domains do not act as point crosslinks, unlike in natural rubber. The physical crosslinks in PU elastomers occupy a substantial volume and also act as fillers. Moreover, the crosslink is not permanent and may deform and degrade under high strain.

Elastomers produced using a two-component molding technology generally show a combination of physical and chemical crosslinks. The chemical crosslinks are introduced by using branched monomers or allophanate formation, which occurs at an index greater than 100. The introduction of chemical crosslinks prevents flow at temperatures where the hard domains begin to melt. Chemically crosslinked PU elastomers, therefore, show improved mechanical stability at higher temperatures. Chemical crosslinks, however, strongly impart the ultimate extensibility of the polymer. The level of chemical crosslinking in PU elastomers is maintained at a sufficiently low level to ensure that the elongation at break is at least 300%. The average molar mass between the chemical crosslinks in cast elastomers typically ranges from approximately 5,000 to 10,000 g/mol.

9.4 Effect of processing on morphology

During the polymerization of PU elastomers, polymerization-induced phase separation will occur. The conversion at phase separation (CPS) depends on the system composition and the reaction conditions. When CPS occurs early, the developed morphology might be coarse, and full cure may not be reached, resulting in an elastomer with weak mechanical properties. In contrast, when the CPS occurs at relatively high conversions, full conversion, and high molecular mass can be expected; however, the phase morphology may be poorly developed. The two phases may be partially mixed, resulting in a polymer exhibiting weak hard domains and, consequently, poor strength and resilience. Therefore, the conversion at which phase separation occurs is critical in achieving the desired mechanical properties. The CPS can be altered by adjusting the monomer choice and system composition, as well as by optimizing the reaction conditions. It should be delayed if the problem is associated with coarse morphology development and insufficient molar mass build-up, or brought forward to earlier conversion to enhance phase separation.

9.4.1 Reaction rate, reaction temperature, and hard block content

The easiest way to tune the CPS in practice is by adjusting the reaction rate, temperature, and hard block content.

Suppose a PU reaction under quasi-isothermal conditions, as it would occur when casting a PU elastomer on a thermostated hot plate. The polymerization process begins after mixing, and the temperature and catalyst concentration determine the reaction rate. The conversion of the monomers to polymers progresses, and the molar mass increases with time. At a given conversion, the hard blocks will have reached a sufficient molar mass to induce phase separation. Fig. 9.13 sketches the conversion with time at two catalyst concentrations, high (the red line at high catalyst concentrations $[cat]_1$) and low (the blue line at low catalyst concentrations $[cat]_2$). The CPS is indicated with a black dot.

Phase separation is a diffusion-controlled process, and the hard blocks need some time for aggregation and domain building. The time required to aggregate must be offset against the reaction rate, whereby an increased reaction rate will delay the CPS. A high catalyst concentration increases the reaction rate, and consequently, the CPS occurs early at relatively high conversions (c_1 at t_1). The reaction rate after phase separation will slow down due to diffusion limitations and reactive group starvation; however, it will likely remain high enough to achieve full conversion with sufficient time. The reaction proceeds more slowly at low catalyst concentrations, and the hard blocks have, on the timescale of the reaction, more time to aggregate. As a result, the phase separation starts later at comparatively low conversions (c_2 at t_2 whereby $t_2 > t_1$ and $c_2 < c_1$). Because the diffusion limitation of the reaction starts at relatively low

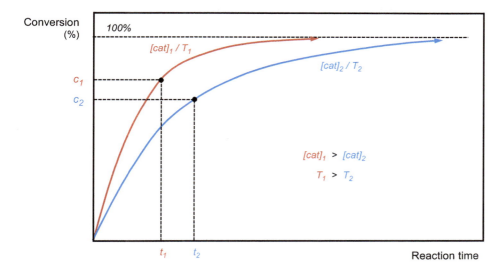

Fig. 9.13: Conversion versus time at different catalyst concentrations ($[cat]_1$ and $[cat]_2$) and, alternatively, at different reaction temperatures (T_1 and T_2).

conversions, it may result in incomplete conversion, coarse morphology development, and poor elastomer properties.

The effect of reaction temperature variation shows some similarity to catalyst variation, as discussed above, and is also illustrated in Fig. 9.13. It now displays the conversion with time at two reaction temperatures: high (represented by the red line at T_1) and low (the blue line at T_2). Phase separation is a diffusion-controlled process that depends on both time and temperature. The temperature increase delays the phase separation process for two reasons. First, the temperature increase causes the reaction to proceed faster, and second, the high reaction temperature keeps the hard blocks in solution for a longer time. The reaction proceeds faster at high temperature (T_1), and the onset of phase separation occurs earlier and at higher conversion (c_1 at t_1). In contrast, at low temperatures (T_2), the onset of phase separation occurs later and at lower conversion. The reaction is likely to reach full conversion at high temperatures; however, there is a risk of incomplete conversion and coarse morphology development at low temperatures.

An earlier CPS at lower reaction temperatures may change the hard block size distribution from the theoretical "most probable" Schulz-Flory distribution to a broader distribution. The broadened size distribution, showing some smaller and longer hard blocks, can affect the crystallization rate and, consequently, the polymer morphology and properties of the final polymer. The longer hard blocks increased the crystallization rate and the extent of crystallization, resulting in the formation of crystalline superstructures [11]. Because the properties of PU elastomers depend on the reaction rate and temperature at which they are produced, they are polymers by process.

The hard block content is an additional variable that can be used to tune the onset of phase separation. An increase in HBC will shift the onset of phase separation to lower conversions because the solubility limit of hard chain segments in the reacting liquid will be reached sooner, and *vice versa*.

9.4.2 Polyol, chain extender, and isocyanate

The CPS depends on the monomers applied; the diisocyanate, chain extender, and polyol structure determine the processing and the final extent of phase separation in the elastomer. Planar and highly symmetric isocyanates and short and even-numbered diol chain extenders enhance phase separation in the final polymer. However, these hard blocks are more prone to phase separation during polymer formation, and the CPS occurs at lower conversions. Similarly, high-molar-mass and apolar polyols result in strong phase separation, yet reaction systems based on such polyols will also phase separate at lower conversions. This highlights a potential conflict in producing high-end elastomers: Formulations based on starting components known for their good elastomer properties may exhibit poor processing, resulting in elastomer formation with inferior or suboptimal properties. The processing of these elastomers can generally be improved by adjusting the reaction conditions. Increased reaction rates and temperatures, and the use of isocyanate prepolymers might alleviate the production problems. Often, a compromise must be found between reliable and reproducible processing and the ultimate mechanical properties. Tab. 9.6 summarizes the effects of the employed reaction components on the conversion at the onset of phase separation.

Tab. 9.6: Influence of the reaction components on the conversion of phase separation.

Reaction component	Resulting in	
	late phase separation	early phase separation
Diisocyanate	Non-planar Non-symmetrical	Planar Symmetrical
Long-chain polyol	Low molecular mass	High molecular mass
	Polar	Apolar
Chain extender	High number of C-atoms	Low number C-atoms
	Odd-numbered	Even-numbered

9.4.3 Final morphology, dependency on time and temperature

Several chemical, physical, and structural parameters control the final polymer morphology and mechanical properties of PU elastomers. The thermodynamic incompatibility of the hard and soft segments drives the phase separation process; however, the kinetics of phase separation determines, in practice, the extent to which phase separation occurs.

PU elastomers will exhibit a phase-separated morphology on the nanometer scale, provided the thermodynamic incompatibility is sufficiently high. Increasing the molar mass as well as a small dispersity index of the hard and soft blocks will enhance phase separation. Hard blocks of planar and symmetric isocyanates and short-chain even-numbered glycols can form strong and highly ordered hard domains, exhibiting aligned hydrogen bonds between adjacent urethane groups. The strong hard segment interaction, in turn, expels polyol from the interphase region, thereby increasing the phase separation. In contrast, competitive intermolecular interaction between the hard and soft blocks will cause phase mixing. Covalent crosslinks will further reduce the system's ability to phase segregate. The extent to which the above structural variables interact determines the polymer morphology and mechanical properties [10].

The development of hard and soft phases, for example, during injection molding of a part, also depends on the reaction conditions employed. The reaction conditions determine the time and temperature the polymer experiences during manufacturing, such as the mold temperature and demolding time. Furthermore, it co-determines the phase separation kinetics governing the phase structure. The cooling rate after demolding and storage conditions of the polymer may further affect the morphology. The phase ripening process at all these stages occurs in a high-viscosity environment. Hence, reaching the equilibrium morphology may take a considerable amount of time. Equilibrium phase morphologies and the ultimate mechanical properties are generally not obtained in commercially produced PU elastomers.

Annealing can improve the mechanical properties of PU elastomers. The process involves heating the polymer to a temperature near the melting point of the hard domains for a prolonged period, typically 20 h at 100 °C for PU elastomers. This facilitates the movement of polymer chains, allowing the hard blocks to align and pack orderly. The driving force is the reduction in free energy gained by lowering the surface area of the relatively wide and diffuse hard domains, which become more compact and thicker. The thickening of the phases only occurs close to the melting temperature of the hard phases when sufficient thermal energy is available, allowing the required molecular motion.

Annealing enhances the ordering of the hard domains, leading to an increase in the urethane-urethane hydrogen bonding interaction, as indicated by changes in the peak positions and shapes of the (N-H) and (C=O) absorptions in IR spectroscopy. Concomitantly, the amount of interphase material is reduced, and the volume fraction of the soft phase increases. Fig. 9.14 illustrates the modulus-temperature behavior before and after annealing. The thickening of the soft phase results in a sharper and lower glass transition, indicating improved low-temperature flexibility. The melting of the hard domains is shifted to higher temperatures, extending the plateau modulus. The crossover temperature depends on the polymer and typically ranges from 40 to 100 °C. After annealing, the sharper and lower glass transition reduces damping (reduced loss factor) at room temperature and improves resilience.

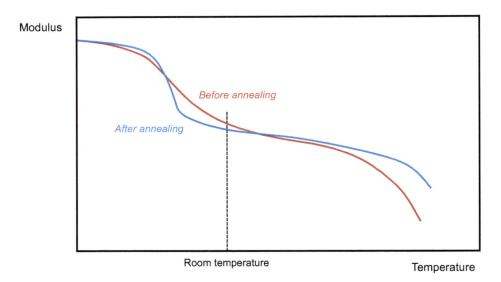

Fig. 9.14: DMA curves of a PU elastomer before and after annealing.

9.5 Performance-related tests

Most standard mechanical properties tests of polymers can also be applied to PU elastomers. For several tests, a distinction is made between compact and foamed material. The most relevant tests are density, hardness, tensile strength, tear strength, compression set, and rebound.

The measurement of the density of solid materials is performed according to ISO 1183. A test specimen is first weighed in air and then in a liquid with a well-known density. The density of the material can be calculated using the buoyancy of the test specimen. The apparent overall and core densities of integral foams can be determined using ISO 845.

The standard hardness testing method (ISO 868) utilizes a durometer to measure the indentation resistance of the elastomer after a force has been applied to a conical indenter. The shape of the indenter varies depending on the hardness of the material being tested. A durometer A (Shore A hardness) features an indenter with a blunted tip and is suitable for softer elastomers, whereas a durometer D (Shore D hardness) has a pointed tip and is suitable for harder materials. The hardness values range from 0 (full penetration) to 100 (no penetration). A related test, Asker C, may be used for characterizing soft elastomer foams. The compression hardness of foams (ISO 3386, Parts 1 and 2 for densities up to and above 250 kg/m^3, respectively) is recorded at compressions of 25, 40, 50, and 65% and reported in kPa.

The tensile tests of compact and flexible cellular elastomers are performed according to ISO 527 and ISO 1798, respectively. A dumbbell-shaped test specimen is extended at a constant rate until it breaks. The recorded stress-strain curve provides the tensile strength and elongation at break, and, if applicable, the stress and elongation at the yield stress, and the measured stress at several preselected elongations. The strength values and moduli are reported in Pascal (kPa or MPa), whereas the elongation at yield and break are reported as a percentage.

The tear strength (ISO 34) indicates how well a material can withstand tearing and is determined by measuring the force required for crack propagation. Force is applied to test samples with a nick in the tensile direction; the tear strength is then calculated and reported in Newton per meter (N/m). Three different methods for specimen shape are specified: trouser, angle, and crescent. The first two are used for foamed elastomers, while the latter is used for compact materials.

Resilience can be determined using two different experimental methods: the pendulum and the ball rebound test. The results provide information on how compact elastomers behave under impact. Using the Schob pendulum rebound resilience test (ISO 4662), the resilience is determined by a freely falling pendulum hammer that is dropped from a specified height and impacts a test specimen with a predetermined amount of energy. The energy stored in the polymer is returned to the pendulum, and its resilience can be calculated from its rebound. The Bayshore Resilience (ASTM D2632) is determined by dropping a ball onto the test specimen and then measuring the rebound height to the initial height ratio. A ratio of 100% indicates a completely elastic response, whereas a ratio of 0% indicates complete energy absorption.

The compression set of elastomers (ISO 815) measures the ability of elastomers to retain their elastic recovery properties at specified temperatures after prolonged compression at a constant strain. The sample is compressed, typically by 25% of its initial height, and exposed to defined temperatures for specified times (often 23 and 70 °C, for 24 h) in its compressed state. It is then released and cooled to room temperature, and its height is measured at the new position. The compression set describes the percentage of the initially applied compression that remains after the material has relaxed.

PU elastomers are used in various specific applications. Performance-related tests, such as flex fatigue, abrasion resistance, and environmental exposure tests, may be required.

9.6 Mechanical properties

Polyurethane elastomers are high-quality engineering materials combining high elasticity with adjustable hardness and low wear. The elastomers exhibit elastic properties over a wide temperature range, typically between 0 and 100 °C, and demonstrate extensibilities upon deformation of at least 300%. The hardness can be adjusted over a wide range, from soft to hard, with values ranging from 30 Shore A to 75 Shore D. Furthermore, high mechanical strength, toughness, and abrasion resistance can be achieved. PU elastomers show good resistance to most non-polar organic solvents such as petrol and gasoline.

PU elastomers also have some disadvantages. The urethane and especially the ester bonds limit their hydrolysis resistance; they are sensitive to strong acids and bases, oxidizing agents, and polar solvents. The reversibility of the urethane bond, which begins at approximately 150 °C, limits the high-temperature resistance of PU elastomers to this temperature range.

9.6.1 Elasticity

PU elastomers can be stretched several times their length, and once the force is removed, the samples return to their original size. The soft phase consists of flexible nonpolar chains with a low T_g fixed by hard domains and chemical crosslinks. The soft phase behaves essentially as an "entropy spring". The chains uncoil when the materials are submitted to strain, reducing their conformational entropy. The restoring force in the stretched sample is caused by the extended chains wanting to regain their original high entropy level.

PU elastomers are viscoelastic materials. Rebound resilience is used to assess the elastic behavior of elastomers when subjected to an impact load. Rebound resilience can be measured using a dropping ball or pendulum method. The test measures the amount of kinetic energy an elastomer sample returns after impact and is expressed as an inverse measure of damping. A perfectly elastic material will return all the energy it has stored during recovery. A viscoelastic material will return less energy because part of the mechanical energy input is lost as heat. The dissipated energy equals the difference in kinetic energy just before and after the collision. The heat formation is caused by internal molecular friction within the polymer during deformation. It can generally be said that in a series of PU elastomers with increasing HBC and concomitantly increasing hardness, the resilience decreases.

A correlation was found between the loss factor (*tan δ*) and Bayshore Rebound for a series of PU elastomers with a broad hardness range, from 55 Shore A to 75 Shore D. The following correlation was found:

$$Rebound = e^{-n \cdot tan\delta}$$

The dynamic mechanical analysis was performed at 110 Hz. In this work, n was about 8 [12].

9.6.2 Strain-stress behavior

PU elastomers exhibit high strength properties, characterized by a high elongation at break and high tensile strength. The hard domains play the most important role in the overall strength properties of an elastomer. The hard domains predominantly strengthen the elastomer by providing hardness to the polymer and plastic deformation under strain. The soft domains contribute to the strength by developing high deformation levels and strain-induced crystallization.

Elastomers based on crystallizable polyols that crystallize upon stretching generate crystalline polyol domains. Plastic deformation of the crystalline domains at high strain results in energy dissipation, which is reflected in the strength and toughness. When soft segment crystallization occurs upon deformation, an upturn in stress is observed, typically occurring at deformations above 200%. Polyester-based elastomers show generally higher tensile strength properties than polypropylene oxide-based elastomers with the same hard block monomers and polymer hardness. The polyester phase exhibits significant intermolecular interactions, and depending on its molecular structure, it can crystallize under strain. These properties are absent in the propylene oxide-based soft phase, and their elastomers are generally weaker. Good mechanical properties can be achieved when using PTMEG polyols. The PTMEG phases exhibit little intermolecular interaction; however, upon strain, they can undergo crystallization.

Hard domains are effective reinforcing agents, provided their volume fraction is above 20%. Below this value, the hard domains are too small to withstand strain, and the polymer has little strength. The intramolecular interaction between the hard blocks in the hard domains is high, and the hard domains can dissipate significant amounts of energy during deformation. The hard domain volume fraction of a series of MDI-based TPUs was investigated as a function of strain [1].

When the volume fraction of hard domains was normalized to one at zero strain, all the TPUs exhibited similar behavior with strain [Fig. 9.15]. The hard domain volume increased during extension to about 50% elongation. The increase in hard domain volume, ranging from 10 to 20% of the original value, was attributed to strain-induced crystallization, thereby orienting the hard blocks in the stress field. After that, all materials showed a linear decrease in the hard-phase volume fraction, reaching zero at about 550% strain, which agreed reasonably well with the elongation-at-break of the TPUs. During this phase, the hard blocks are gradually pulled out of the hard domains with increasing strain until no hard phase remains. This suggests that

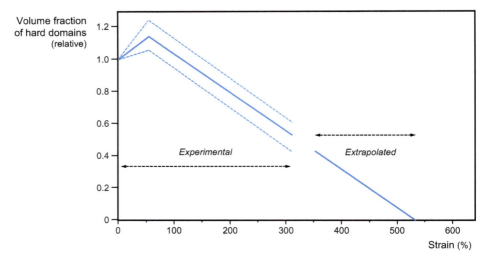

Fig. 9.15: The hard domain volume fraction versus strain. Data were recorded till 300% strain and extrapolated to zero hard domain content. The drawn line shows the average value of all the TPUs; the dotted lines represent the deviation (adapted from [1] with the kind permission of John Wiley and Sons).

the materials can be extended as long as hard domains are present, allowing them to manage the strain, and rupture when all domains are sacrificed.

9.6.3 Effect of hard block content on properties

The morphology of segmented polyurethanes largely depends on their hard block content. The hard phases are dispersed in the continuous polyol phase until an HBC of approximately 40% [Fig. 9.16a]. With increasing HBC from 40 to 60%, the morphology changes to a bi-continuous morphology of hard and soft domains [Fig. 9.16b]. Phase inversion occurs at HBCs above 60%, where the soft domains are dispersed throughout a continuous hard phase [Fig. 9.16c].

With increasing HBC, the hard domains become thicker and stronger, increasing the strength of the elastomers. Superior resilience values are achieved at HBCs between 20 and 40%, with the best tensile strength values between 40 and 60%. When the HBC is further increased, the system phase inverts and the elastomers exhibit reduced elasticity and extensibility, becoming hard, inelastic, and relatively brittle. Therefore, the HBC is between 20 and 50% in most elastomer applications.

The effect of increasing HBC in a series of TPUs based on polyester polyol, 1,4-butanediol, and 4,4'-MDI is shown in Fig. 9.17. An increasing HBC content increases the polymer hardness at use temperature and improves its high-temperature stability. The phase separation, however, becomes less well developed because more interphase

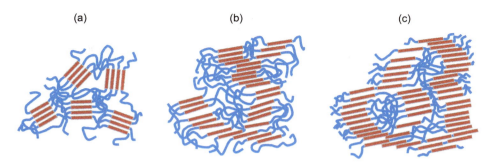

Fig. 9.16: Morphology of segmented PUs with increasing hard block content.

material is formed. The glass transition of the soft phase is broadened and shifted to higher temperatures, and the modulus plateau is less well developed [13].

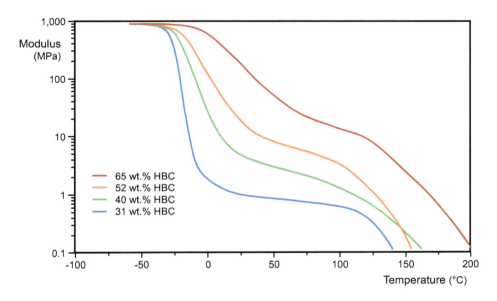

Fig. 9.17: Effect of HBC on shear modulus in a solid PU elastomer adopted from [13] (with the kind permission of Carl Hanser Verlag).

Hard domains are effective reinforcing agents; they are covalently bound and deformable. Plastic deformation of the hard domains reduces stress concentrations and delays fracture, thereby increasing the strength and toughness of the elastomers. The strength and extensibility of a series of PU elastomers with increasing HBCs were compared. The elastomers were prepared from a polypropylene oxide-based triol polyol, 1,4-butanediol, and 4,4'-MDI, whereby the HBC was increased from 11 to 32 wt.% [Fig. 9.18]. With increasing HBC, the polymer becomes more rigid and stronger. The polymer can withstand more stress, exhibiting higher elongation and tensile strength at break. The tensile strength and elongation at break, however, exhibit a semi-circular behavior; at high HBC, the elongation at break may decrease again because the polymer extensibility decreases.

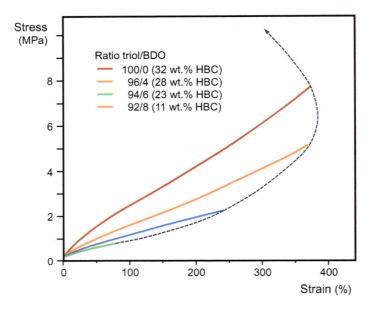

Fig. 9.18: Stress-strain behavior of a PU elastomer based on triol polyol, 1,4-butanediol, and 4,4′-MDI at various HBCs (wt.%).

9.7 Applications

9.7.1 TPU

TPUs can be produced in discontinuous batch processes; however, nowadays, they are mainly produced continuously by reactive extrusion or band casting. TPUs are available as chips or pellets, which can be processed using conventional thermoplastic technologies, such as injection molding, extrusion, blow molding, and calendaring. Due to its excellent flexibility, abrasion resistance, and oil resistance, TPUs are utilized in various applications, including footwear, cable, hose, membrane, sheeting, and various automotive applications [Fig. 9.19].

Most TPUs are based on 4,4′-MDI as the isocyanate, PTMEG or adipate-based polyester as the polyol, and 1,4-butanediol as the chain extender. Aliphatic isocyanates are used when light stability is required, for instance, in film applications. The polymers differ in

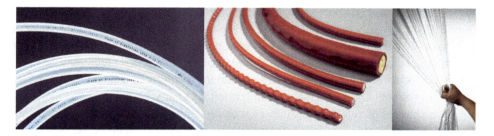

Fig. 9.19: Typical TPU applications, including hoses, cable protection, and foils (with the kind permission of BASF SE).

HBC, which, in turn, determines their hardness. TPUs with hardnesses ranging from 50 Shore A to 75 Shore D can be produced.

Using the band-casting process the starting materials are fed into a mixing head and poured onto a heated conveyor belt. The liquid mixture reacts and solidifies. The polymer sheet is removed from the belt at the conveyor's end and ground into chips. The chips can, if required, be pelletized. The chips are then fed into an extruder, melted, extruded through a nozzle, and cut into granules.

Using reactive extrusion, the preheated starting materials are fed into a twin-screw extruder, which is heated to a temperature above the melting point of the TPU. After the reaction is completed, the melt is pelletized.

9.7.2 Cast elastomers

Cast elastomers are produced by mixing two liquid components and pouring the mixture into a heated mold, where it reacts to form a compact polymer. The two components can be a mixture of polyol and chain extender with isocyanate (one-shot technology); however, in most cases, prepolymer technology is applied whereby part (semi or quasi-prepolymer technology) or all the polyol (full prepolymer technology) has been pre-reacted with isocyanate. Bubble formation during the reaction is to be avoided. Both components are, therefore, degassed by vacuum treatment. Additionally, desiccants such as zeolite paste may be added to the A-component to absorb residual moisture. The systems usually require long open times and fast demolding times. This can be achieved by using delayed-action catalysts that require some temperature activation to kick in.

Two processing technologies are employed: cold casting and hot casting. The hot cast technology generally produces elastomers with superior properties; however, the cold cast technology has superseded the hot cast technology in many applications because it is easier to process.

The cold-cure process utilizes low-viscosity reaction components that can be processed below 50 °C. Compared to hot cure systems, cold cure systems react faster and do not require thermal post-treatment. The polyisocyanate of choice is uretonimine-modified or prepolymerized 4,4'-MDI. The polyols can be selected from PTMEG, molded polyether, or adipate-based polyester polyols. The most used chain extender is 1,4-butanediol; however, other glycols and diamine chain extenders may be used. The systems typically contain branched monomers, resulting in wide-meshed polymer networks with enhanced thermal and dimensional stability.

Typical uses of cold-cast elastomers include fenders for cars and boats, wheels, rollers for kickboards and inline skates, sieves, vibration-dampening elements, and protective edgings, such as those used in furniture applications. A medical application of cold-cast elastomers is the sealing of dialysis filters.

Hot-cure cast elastomers are produced between 60 and 130 °C and require thermal treatment after demolding to complete curing and morphology development. The high temperatures allow the processing of viscous polyols and low-melting solid compounds. The crosslinking in hot-cure elastomers is generally lower than that of cold-cure elastomers. The preferred isocyanates are MDI, TDI, or NDI.

NDI-based elastomers show excellent mechanical properties. NDI-based cast polyurethane elastomers can have a solid or a cellular structure. These high-performance elastomers are produced using full prepolymer technology. First, a prepolymer of NDI (melting point 127 °C) and the polyester polyol is prepared. The prepolymer is mixed with glycols or water in the second step and poured into open molds. The thermal stability of NDI-based prepolymers is limited, and they are prepared just before use. Glycol-extended solid elastomers are produced at temperatures of 100 °C and above, whereas water-blown urea-extended cellular elastomers require approximately 90 °C. After demolding, the elastomers undergo thermal treatment to achieve their final mechanical and dynamic properties. Solid NDI elastomers cover a hardness range from approximately 65 Shore A to 60 Shore D and are used to produce wheels and castors that can withstand high dynamic loads. Microcellular NDI elastomers, spanning a density range of 300 – 800 kg/m^3, exhibit excellent damping and combine high-volume compressibility with minimal transverse expansion. The foamed elastomers are used as auxiliary springs, NVH (noise, vibration, harshness) elements, mounting pads, and crane bumpers.

9.7.3 Microcellular elastomers

Microcellular elastomers are molded high-density elastomer foams with overall densities ranging from 200 to 800 kg/m^3. Their cells are much finer than those of low-density foams. At densities above 350 kg/m^3, the foams are closed-cell, and the cells are isolated from one another; at lower densities, the cells may interconnect, increasing the openness of the foam. Foaming is obtained by adding water or a combination of water and a physical blowing agent. As a result of the molding process, a skin is formed on the surface of the foam. The skin is an integral part of the foam; therefore, the foam is referred to as integral skin foam. Fig. 9.20 illustrates a schematic density distribution in an integral foam from its surface to the core. The skin thickness and surface definition improve when using a physical blowing agent.

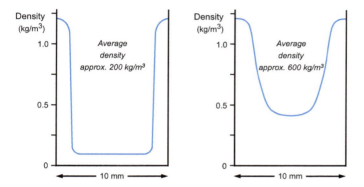

Fig. 9.20: Density distribution in integral skin foams.

When the reaction mixture is injected into the mold and starts to react, carbon dioxide is formed, and heat is generated. The physical blowing agent evaporates, the foam expands, and the pressure in the mold increases. The free rise density of the foam is low, allowing significant overpacking of the mold. The foam core may reach temperatures of approximately 150 °C, whereas the material at the mold surface will only reach the mold temperature, which is maintained at around 50 °C. The mold pressure and temperature are chosen so the physical blowing agent at the mold surface does not evaporate, allowing a dense skin to form. Purely water-driven systems can also provide adequate skin buildup; by system and process optimization, the skin of the foam may be sufficient to meet the required application demands. PU shoe soles, for example, exhibit skin formation without using physical blowing agents. Other examples of integral skin foams are steering wheels, gearshift knobs, armrests, and bicycle saddles, expanded with water or the combination of water and physical blowing agent.

9.7.4 Other elastomer applications

RIM elastomers are produced by injecting a reactive mixture into a mold, short Reaction Injection Molding. The polymer articles exhibit densities above 900 kg/m^3 and are, therefore, virtually compact. Due to the high system reactivity and the relatively large sizes of the produced parts, high-pressure mixing is necessary. The reactivity increases with the use of more reactive amine-based monomers, from conventional RIM (molded polyol/short chain glycol) to High-Speed HS-RIM (reactive polyol/DEDTA) to Polyurea-RIM (polyether amine/DETDA). RIM parts are often reinforced using filler materials. Reinforced R-RIM contains milled glass or short mineral fibers, whereas Structural S-RIM is reinforced with glass-fiber mats. These RIM technologies produce highly dimensional stable articles for the automotive industry, such as body panels, bumpers, spare wheel covers, and truck floors.

Elastomers that are spray applied to cover a surface are called spray elastomers or spray coatings. The elastomer coating systems are solvent free and applied at thicknesses ranging from 0.5 to 10 mm and typically utilize MDI prepolymer technology. Special two-component high-pressure spray machines have been developed to apply spray coats on-site. Most spray systems are glycol-based (molded polyol/short chain diol) and applied as a protective coating in liner applications, such as tank, container, and pipeline coatings. Fast-curing polyurea (polyether amine/DEDTA) systems are used in applications where post-cure is impossible, such as roofing, flooring, and secondary containment.

Artificial leather is produced using a coagulation process by treating a TPU solution with water. The TPU, dissolved in a good solvent such as dimethylformamide, is applied to a fleece or textile substrate. The freshly coated substrate is then pulled through several solvent-water baths with increasing water content, thus stepwise reducing the solubility of the TPU. The wet TPU coagulate is dried, yielding a breathable artificial textile leather sheet.

Elastomeric fibers are produced using dry and wet spinning techniques. First, a TPU is dissolved or prepared in a solvent. Using the wet spinning process, the solution is spun into a non-solvent, e.g., water, to form the TPU filament. The dry spinning process also uses a TPU solution, which is spun into hot air, evaporating the solvent and yielding the fiber. The highly elastic fibers can be combined with natural or synthetic fibers to produce elastic woven textiles such as socks, hoses, underwear, and swimsuits.

References

Further reading

– R. Leppkes, Polyurethanes, Süddeutscher Verlag Onpact GmbH, Munich, 2012, ISBN 978-3-86236-039-0.
– C. Prisacariu, Polyurethane Elastomers, 2011, Springer Verlag Wien, ISBN 978-3-7091-0513-9.

References

[1] A. Stribeck, X. Li, A. Zeinolebadi, E. Pöselt, B. Eling, S. Funari, Morphological changes under strain for different Thermoplastic Polyurethanes monitored by SAXS related to strain at break, Macromol. Chem. Phys., 2015, 216, 2318–2330.
[2] A. Stribeck, E. Pöselt, B. Eling, F. Jokari-Sheshdeha, A. Hoell, Thermoplastic polyurethanes with varying hard-segment components. Mechanical performance and a filler-crosslink conversion of hard domains as monitored by SAXS, Eur. Polym. J., 94 (2017) 340–353.

[3] Z.S. Petrovic, J. Ferguson, Urethane elastomers, Progr. Polym. Sci., 16 (1991) 695–836.

[4] C. Zhang, J. Hu, S. Chen, J. Fenglong, Theoretical study of hydrogen bonding interactions on MDI-based polyurethane, J. Mol. Model., 16, 1391–1399 (2010).

[5] P. Mackey, N. Limerkens, V. Kapasi, B. Fogg, The Polyurethanes Book, D. Randall and S. Lee, Eds, Wiley, 2002, ISBN 0-470-85041-8.

[6] J.E. Moussa, J.J.P. Stewart, MOPAC software version 22.1.0, doi: 10.5281/zenodo.6511958, released 2023-09-25.

[7] L. Born, H. Hespe, J. Crone, K.H. Wolf, The physical X-ray investigation of model compounds, Colloid Polym. Sci., 260(9), 818 (1982).

[8] R.M. Versteegen, R.P. Sijbesma, E.W. Meijer, [n]-Polyurethanes: Synthesis and characterization, Angew. Chem. Int. Ed., 1999, 38 (19), 2917–2919.

[9] J. Blackwell, K.H. Gardner, Structure of the hard segments in polyurethane elastomers, Polymer, 20 (1), 16 (1979).

[10] I. Yilgör, E. Yilgör, G.L. Wilkes, Critical parameters in designing segmented polyurethanes and their effect on morphology and properties: A comprehensive review, Polymer, 58 (2015) A1eA36.

[11] Z. Gao, Z. Wang, Z. Liu, L. Fu, X. Li, B. Eling, E. Pöselt, E. Schander, Z. Wang, Hard block length distribution of thermoplastic polyurethane determined by polymerization-induced phase separation, Polymer, 256 (2022) 125236.

[12] B.L. Williams, Correlation between tan δ and Bashore Rebound for polyurethane elastomers, J. Appl. Polym. Sci., 16 (1972) 3387–3388.

[13] G. Oertel, Polyurethane, Kunststoff Handbuch (Becker/Braun), Band 7, 1993, 47, ISBN 3-446-16263-1.

10 Sustainability and outlook

Sustainability will become one of the main drivers for the future development of polyurethanes, and environmental issues will become more critical. This includes the raw materials and additives used in the production and the handling of end-of-life PU materials.

Polyols with an improved carbon footprint can be produced, which, in addition, may allow the development of polyurethanes with improved properties. Emissions and odors are growing concerns in building and automotive interior applications, and low-emission and low-odor PU foam systems are under development. Utilizing novel, environmentally benign blowing agents in rigid foam may enhance insulation properties. End-of-life polyurethanes can be physically and chemically recycled into reconstituted foams and starting materials for future applications.

This chapter is a shortened version of reference [1].

10.1 Improved carbon footprint of PU starting materials

Increasing concern over environmental pollution, climate change caused by greenhouse gases, and future petroleum scarcity has triggered significant efforts to develop novel and sustainable polyols and isocyanates from renewable resources, such as naturally occurring vegetable oils and industrial biomasses.

The intensive research activities on the transformation of carbohydrate-containing biomasses (sugar, starch, and lignocellulose) have led to the successful large-scale production of various bio-based monomers, such as ethylene glycol, 1,3-propanediol, 1,4-butanediol, isosorbide, succinic acid, adipic acid, azelaic acid, sebacic acid, terephthalic acid, 1,5-pentamethylenediamine, and 1,6-hexamethylenediamine.

The diols 1,3-propanediol and 1,4-butanediol produced from sugar fermentation processes are used to produce bio-based polyether polyols such as poly(trimethylene ether) glycol (Cerenol®, DuPont) and polytetrahydrofuran (PolyTHF® 1,000, BASF). Furthermore, bio-based diols are combined with bio-based dicarboxylic acids to produce polyester polyols. These polyester polyols serve as drop-in replacements for petroleum-based polyester polyols in PU elastomer and TPU applications.

Bio-based pentamethylene diisocyanate is produced in the reaction of phosgene and 1,5-pentamethylene diamine obtained by sugar fermentation (Desmodur® eco N 7300 and STABiO®, by Covestro and Mitsui, respectively). Recently, Covestro developed a process to produce aniline from biomass. Sugar is converted through fermentation into an intermediate product, which is then catalytically converted to aniline in a second step. The aniline can then be used in the manufacturing of MDI.

Triglycerides can be hydrolyzed to produce glycerine and fatty acids. The fatty acids can be incorporated into polyester or hybrid polyester-ether polyols. Unsaturated

https://doi.org/10.1515/9783110744583-010

oils, such as rapeseed, soybean, and palm kernel, serve as raw materials for the production of natural oil polyols. Various reactions can be performed with the double bonds, such as epoxidation, hydroformylation, and ozonolysis. Most natural oil polyols are synthesized from epoxidized vegetable oils. The epoxidized oils are reacted with, for example, alkanolamines, water, or mono-functional alcohols to give the polyol. Several natural oil-based polyols are commercially available, including Sovermol® (BASF), BiOH® (Cargill), and Renuva® (Dow). Sovermol® products are designed for coating and adhesive applications, whereas BiOH® and Renuva® polyols are developed for flexible foam applications.

Unsaturated fatty acids can be dimerized through their double bonds to obtain dimer acids, which can be reduced to dimer diols (e.g., Pripol®, Cargill; Sovermol®, BASF). These hydrophobic polyols are used in elastomer, coating, and adhesive applications. Dimer fatty acid is a precursor for producing dimer fatty acid diisocyanate (DDI®, BASF).

Castor oil contains naturally occurring secondary hydroxyl groups and is used in coating and adhesive applications. Castor oil can be further alkoxylated with propylene and ethylene oxide to yield polyols for flexible slabstock foam applications (Lupranol® Balance 50, BASF).

Aramco (Converge®) and Covestro (Cardyon®) have recently developed processes for producing polyether-carbonate polyols from propylene oxide and carbon dioxide. The carbonate polyols are used in flexible and rigid foams, coatings, adhesives, and elastomers.

10.2 Emission and odor

PU indoor application materials must meet stringent emission and odor standards. Automotive OEM requirements for indoor air quality have made the reduction of volatile organic compounds (VOCs) and aldehyde emissions increasingly important to the PU industry. In particular, formaldehyde and acetaldehyde are of concern because they are classified as possible human carcinogens.

Developing low-emission and low-odor PU foams requires careful selection of raw materials and additives specifically designed to control and minimize the presence of volatile by-products. Suppliers of additives, polyols, and isocyanates have intensified their efforts to minimize the odor and emissions associated with their starting materials. One of the primary sources of VOCs and odors in open-cell PU foams is the use of low-molecular-mass tertiary amine catalysts. Molded polyols with built-in catalytic activity (e.g., Voractive® polyols, DOW) can significantly reduce the amount of amine catalysts required for the reaction. Furthermore, amine catalysts containing isocyanate-reactive groups such as hydroxyl or amino groups were developed. The reactive groups ensure the immobilization of the catalyst within the final polymer network and prevent its release.

Furthermore, research efforts are focused on reducing aldehyde emissions and odor by developing and incorporating efficient aldehyde scavengers. Combining improved raw materials and novel additives enables the design of PU materials that meet the most stringent standards for odor and emissions.

10.3 Insulation

Closed-cell rigid PU foam is known for its excellent insulation properties and is used in appliance and construction applications. Legal pressure on energy consumption standards creates a demand for further improved products.

The foam density, cell size, and thermal conductivity of the cell gas determine the thermal conductivity of a PU foam. The latter gives the largest contribution to the thermal conductivity of the foam. Recently, hydrofluoroolefins (HFO) were introduced. These novel environmentally benign blowing agents are non-flammable and exhibit low thermal conductivities (e.g., Opteon® 1100, Chemours). With current production technologies, foams with cell sizes of approximately 150 μm can be produced, and in conjunction with HFO blowing agents, insulation values of 16 – 18 mW/(m·K) can be achieved. During their lifetime, the rigid foams will save more than 100 times the energy required to produce the insulation material.

Although rigid PU foams are generally closed-cell, it is also possible to produce open-cell rigid foams. Such foams can be used as the support medium in Vacuum Insulation Panels (VIPs). At pressures below 0.001 kPa the overall thermal conductivity can reach values as low as 5 – 9 mW/(m·K). VIPs are used in premium packaging and appliance applications with the lowest energy consumption ratings.

10.4 Recycling of PU

The management of plastic waste has become a major global challenge. In principle, the following recovery options are possible: mechanical recycling, monomer recovery, feedstock recycling, incineration, and landfill.

The handling of end-of-life polyurethane materials is expected to undergo significant changes in the coming years. The most important change will be to divert significant amounts of PU waste from landfills. While large amounts will need to be incinerated in the foreseeable future, various initiatives aim to find alternative methods for treating production and post-consumer PU waste.

The foamed nature, thermoset character, and the variation in molecular composition may pose a challenge in the recycling of post-consumer PU waste, in which, in principle, all types of PU foam may be present. Thermochemical recycling may then be considered. There are three main technologies for thermochemical recycling: pyrolysis, gasification, and hydrogenation. The shared underlying principle of these technologies is transforming plastic waste into gas and oil that can be used as raw material for the chemical industry.

Recovery of foam for PU-specific recycling purposes may be an option for applications where the amount of used PU is substantial and collection and disassembly are feasible, such as flexible foam from mattresses. The PU can then be recycled in two ways: mechanically, where the material is reused in its polymer form, and chemically, which, in principle, breaks down the material into its original building blocks or precursors.

Mechanical recycling of PU foam waste involves three steps. First, the foams have to be ground to a specified particle size. Subsequently, the foam particles are wetted with a binder, then pressed and heated, yielding the final product. Mechanical recycling is a low-cost operation that is easy to implement.

Chemical recycling is possible due to the thermally reversible nature of the urethane bond. The urethane can be degraded using organic compounds with active hydrogen groups. Depolymerization is generally catalyzed, and relatively high temperatures are required for the process. The crude reaction product can then be separated into various product streams, notably polyol and aromatic amine from the employed isocyanate. The polyol can be reused in foam production, and the amine can be used as a starter for polyols or reconverted into isocyanate. The main PU recycling processes include acidolysis, glycolysis, hydrolysis, and aminolysis. The most employed processes are acidolysis and glycolysis. Using acidolysis, the foam is reacted with a mixture of carboxylic acid and polyol, yielding reclaimed polyol. Glycolysis involves the chemical decomposition of urethane by short glycols. Sometimes, the glycols are combined with low-molecular-mass amines, such as alkanolamines. The process yields polyol and urethane-containing glycolysis products that can be reused to produce PU.

The development of PUs with novel, cleavable chemical bonds has gained increasing attention. Various cleavable bonds, including acetal, imine, and triazine, have been investigated on a lab scale. Such bonds enable the controlled depolymerization of the material under specific conditions, yielding well-defined molecules that can be separated and purified, thus facilitating closed-loop recycling.

10.5 Epilogue

Utilizing bio-based feedstocks to produce PU starting materials and recycling will be key drivers of innovation in the future. However, traditional aspects such as insulation properties, emissions, flammability, and catalysis must also be addressed.

Starting materials based on renewable resources are not necessarily genuinely sustainable solutions. Renewables are inevitably part of the general food and environmental debate and require proper life-cycle assessments.

The regulatory field for the recovery and recycling of foam waste is developing. Several countries are setting up collection and recycling schemes for flexible foam from mattresses. Two main challenges are the product quality of the recycled monomers and the cost. In the long term, PUs that are more apt for recycling should be developed, for instance, by incorporating novel unzippable linkages.

Improved resistance to fire is required, especially in rigid foam construction applications. Traditionally, halogenated and phosphorus fire retardants were applied, the former being active in the gas phase as a radical quencher and the latter in the solid phase as a charring catalyst. However, the use of halogens is discouraged because of environmental concerns. The current focus is on developing novel phosphorus-based fire retardants that exhibit both solid- and gas-phase activity. PIR foam, fire retarded with phosphorus based compounds, are currently available as halogen free solutions for the building industry.

Novel environmentally benign urethane catalysts are required. Catalysts based on tin and mercury are reasonably stable in polyurethane reaction mixtures, and exhibit high reactivity and selectivity in the urethane reaction. Mercury-based catalysts have been discontinued due to environmental concerns, and tin-based catalysts are being phased out where possible. The industry has been extensively searching for alternative, environmentally benign catalysts, but drop-in solutions have yet to be found. Titanium, zirconium, bismuth alcoholates, acetylacetonates, and carboxylates show high urethane selectivity but suffer from poor blend stability.

References

[1]   B. Eling, Ž. Tomović, V. Schädler, Current and Future Trends in Polyurethanes: An Industrial Perspective, Macromol. Chem. Phys., 2020, 2000114.

11 Symbols and abbreviations

Symbols

γ	Surface tension
δ	Phase angle
ε	Strain
η	Viscosity
λ	Thermal conductivity (λ-value)
ρ	Density
σ	Stress
Φ	Porosity
ω	Angular frequency
A	Area
E	Young's modulus
E'	Storage modulus
E''	Loss modulus
EM	Equivalent mass
F	Free energy
f_n	Number average functionality
G	Shear Modulus
G'	Shear storage modulus
G''	Shear loss modulus
HBC	Hard block content
L	Length, thickness
M	Molecular mass
mp	Melting point
OHv	Hydroxyl value, hydroxyl number
P	Pressure
q	Heat flow
r	Radius
T	Temperature
T_g	Glass transition temperature
T_m	Melting temperature
t	Time
$\tan \delta$	Loss factor

Abbreviations

ABA	Aminobenzylaniline
AFM	Atomic force microscopy
AIBN	2,2-Azo-bis(isobutyronitrile)
APP	Ammonium polyphosphate
ASTM	American Society for Testing and Materials (American industry standard)
BDMAEE	Bis(2-dimethylaminoethyl) ether

https://doi.org/10.1515/9783110744583-011

BS	British standard
CFC	Chlorofluorocarbon
CLD	Compression load deflection
CPS	Conversion at phase separation
CVE	Chemical viscoelastic (foam)
DBTDL	Dibutyltin dilaurate
DEEP	Diethyl ethylphosphonate
DETDA	Diethyl toluene diamine
DIN	"Deutsche Industrie-Norm" (German industry standard)
DIY	Do it yourself
DMA	Dynamic mechanical analysis
DMC	Double metal cyanide (catalyst)
DMCHA	N,N-dimethylcyclohexylamine
DMPA	N,N-dimethylphenylamine
DSC	Differential scanning calorimetry
EA	Energy absorbing (foam)
ECHA	European chemical agency
EN	European standard ("Europäische Norm")
EO	Ethylene oxide
FIGRA	Fire growth rate
FMVSS	Federal motor vehicle safety standard
GC	Gas chromatography
GPC	Gel permeation chromatography
GWP	Greenhouse warming potential
HACS	Humid aged compression set
HBC	Hard block content
HCFC	Hydrochlorofluorocarbon
HCFO	Hydrochlorofluoroolefin
HDI	Hexamethylene diisocyanate
HFC	Hydrofluorocarbon
HFO	Hydrofluoroolefins
HLB	High-load bearing (foam)
$H_{12}MDI$	Hydrogenated MDI (methylene dicyclohexyl diisocyanate)
HR	High resilient (foam)
IMC	In-mold coating
IPDI	Isophorone diisocyanate
IR	Infrared (spectroscopy)
ISO	International organization for standardization
LD	Low-density (foam)
LOI	Limiting oxygen index
MBOCA	4,4'-methylene-bis(2-chloroaniline)
MDA	Methylenedianiline
MDI	Methylene diphenyl diisocyanate
MDU	Dimethyl urethane of MDI
MNDO	Modified neglect of diatomic overlap (semi-empirical quantum calculation method)
MT	Blend of MDI and TDI (main component MDI)
NDI	1,5-Naphthylene diisocyanate
NDU	Dimethyl urethane of NDI
NF	French standard ("Norme Française")

OCF	One component foam
ODP	Ozone depletion potential
OEM	Original equipment manufacturer
p.b.w.	Parts by weight
PDI	Pentamethylene diisocyanate
PE	Polyethylene
PEG	Polyethylene glycol
PET	Polyethylene terephthalate
PHD	Poly(hydrazodicarbonamide) ("Polyharnstoffdispersion")
PIPA	Polyisocyanate polyaddition (polyols)
PIR	Polyisocyanurate
PMDI	Polymeric MDI
PMPO	Polymer-filled polyether polyols
PO	Propylene oxide
PP	Polypropylene
PPDI	p-Phenylene diisocyanate
PPG	Polypropylene glycol
PTMEG	Poly(tetramethylene ether) glycol
PU	Polyurethane
PVC	Polyvinylchloride
PVE	Pneumatic viscoelastic (foam)
R-RIM	Reinforced reaction injection molding
RDP	Resorcinol bis(diphenyl phosphate)
RHL	Reverse heat leakage
RIM	Reaction injection molding
rpm	Rotations per minute
RS	Rectangular shape
S-RIM	Structural reaction injection molding
SAN	Styrene acrylonitrile (polymer)
SAXS	Small-angle X-ray scattering
SBI	Single burning item
SEM	Scanning electron microscopy
SFC	Spray foam can
SMOGRA	Smoke growth rate
SPF	Spray polyurethane foam
SPFA	Spray Polyurethane Foam Alliance
SVHC	Substance of very high concern
TCPP	Tris(2-chloropropyl) phosphate
TDA	Toluene diamine
TDI	Toluene diisocyanate
TDU	Dimethyl urethane of TDI
TEDA	Triethylenediamine; 1,4-diazabicyclo[2.2.2]octane
TEM	Transmission electron microscopy
TEP	Triethyl phosphate
TF	Thermo-formable (foam)
THR	Total heat release
TM	Blend of TDI and MDI (main component TDI)
TMXDI	m-Tetramethylxylylene diisocyanate
TODI	o-Tolidine diisocyanate (3,3'-dimethyl-4,4'-biphenylene diisocyanate)

TODU	Dimethyl urethane of TODI
TPP	Triphenyl phosphate
TPU	Thermoplastic polyurethane
TTT	Time-temperature-transformation (diagram)
UL	Underwriters laboratories
USA	United States of America
UV	Ultraviolet (light)
VIP	Vacuum insulation panel
vol.%	Percent by volume
VOC	Volatile organic compound
WAXS	Wide-angle X-ray scattering
wt.%	Percent by weight

12 Unit conversion tables

Length units

Millimeters	Centimeters	Meters	Inches	Feet
mm	cm	m	in	ft
1	$1.0000 \cdot 10^{-1}$	$1.0000 \cdot 10^{-3}$	$3.9370 \cdot 10^{-2}$	$3.2808 \cdot 10^{-3}$
$1.0000 \cdot 10^{1}$	1	$1.0000 \cdot 10^{-2}$	$3.9370 \cdot 10^{-1}$	$3.2808 \cdot 10^{-2}$
$1.0000 \cdot 10^{3}$	$1.0000 \cdot 10^{2}$	1	$3.9370 \cdot 10^{1}$	$3.2808 \cdot 10^{0}$
$2.5400 \cdot 10^{1}$	$2.5400 \cdot 10^{0}$	$2.5400 \cdot 10^{-2}$	1	$8.3333 \cdot 10^{-2}$
$3.0488 \cdot 10^{2}$	$3.0488 \cdot 10^{1}$	$3.0488 \cdot 10^{-1}$	$1.2000 \cdot 10^{1}$	1

Area units

Millimeter square	Centimeter square	Meter square	Inches square	Foot square
mm^2	cm^2	m^2	in^2	ft^2
1	$1.0000 \cdot 10^{-2}$	$1.0000 \cdot 10^{-6}$	$1.5500 \cdot 10^{-3}$	$1.0764 \cdot 10^{-5}$
$1.0000 \cdot 10^{2}$	1	$1.0000 \cdot 10^{-4}$	$1.5500 \cdot 10^{-1}$	$1.0764 \cdot 10^{-3}$
$1.0000 \cdot 10^{6}$	$1.0000 \cdot 10^{4}$	1	$1.5500 \cdot 10^{-3}$	$1.0764 \cdot 10^{1}$
$6.4516 \cdot 10^{2}$	$6.4516 \cdot 10^{0}$	$6.4516 \cdot 10^{-4}$	1	$6.9444 \cdot 10^{-3}$
$9.2903 \cdot 10^{4}$	$9.2903 \cdot 10^{2}$	$9.2903 \cdot 10^{-2}$	$1.4400 \cdot 10^{2}$	1

Volume units

Centimeter cube	Meter cube	Liter	Inches cube	Foot cube
cm^3	m^3	L	in^3	ft^3
1	$1.0000 \cdot 10^{-6}$	$1.0000 \cdot 10^{-3}$	$6.1024 \cdot 10^{-2}$	$3.5315 \cdot 10^{-5}$
$1.0000 \cdot 10^{6}$	1	$1.0000 \cdot 10^{3}$	$6.1024 \cdot 10^{4}$	$3.5315 \cdot 10^{1}$
$1.0000 \cdot 10^{3}$	$1.0000 \cdot 10^{-3}$	1	$6.1024 \cdot 10^{1}$	$3.5315 \cdot 10^{-2}$
$1.6387 \cdot 10^{1}$	$1.6387 \cdot 10^{-5}$	$1.6387 \cdot 10^{-2}$	1	$0.5787 \cdot 10^{-4}$
$2.8317 \cdot 10^{4}$	$2.8317 \cdot 10^{-2}$	$2.8317 \cdot 10^{1}$	$1.7280 \cdot 10^{3}$	1

https://doi.org/10.1515/9783110744583-012

Mass units

Grams	Kilograms	Metric tons	Pounds
g	kg	t	lb
1	$1.0000 \cdot 10^{-3}$	$1.0000 \cdot 10^{-6}$	$2.2046 \cdot 10^{-3}$
$1.0000 \cdot 10^{3}$	1	$1.0000 \cdot 10^{-3}$	$2.2046 \cdot 10^{0}$
$1.0000 \cdot 10^{6}$	$1.0000 \cdot 10^{3}$	1	$2.2046 \cdot 10^{3}$
$4.5359 \cdot 10^{2}$	$4.5359 \cdot 10^{-1}$	$4.5359 \cdot 10^{-4}$	1

Density units

Gram/liter	Kilogram/meter cube	Pound/foot cube
g/L	$1 \ kg/m^{3}$	lb/ft^{3}
1	1	$6.2428 \cdot 10^{-2}$
$1.6019 \cdot 10^{1}$	$1.6019 \cdot 10^{1}$	1

Pressure units

Bar	Kilopascal	Millimeter of mercury	Pound/inch square
bar	kPa	mm Hg	lb/ft^{3} or psi
1	$1.0000 \cdot 10^{2}$	$7.5001 \cdot 10^{2}$	$1.4504 \cdot 10^{1}$
$1.0000 \cdot 10^{-2}$	1	$7.5001 \cdot 10^{0}$	$1.4504 \cdot 10^{-1}$
$1.3332 \cdot 10^{-3}$	$1.3332 \cdot 10^{-1}$	1	$1.9337 \cdot 10^{-2}$
$6.8948 \cdot 10^{-2}$	$6.8948 \cdot 10^{0}$	$5.1715 \cdot 10^{1}$	1

Temperature

Conversion	Celsius	Fahrenheit	Kelvin
Celsius to	–	$T(°F) = \dfrac{9}{5} \cdot T(°C) + 32$	$T(K) = T(°C) + 273.15$
Fahrenheit to	$T(°C) = \dfrac{5}{9} \cdot (T(°F) - 32)$	–	$T(K) = \dfrac{5}{9} \cdot (T(°F) + 459.67)$
Kelvin to	$T(°C) = T(K) - 273.15$	$T(°F) = \dfrac{9}{5} \cdot T(K) - 459.67$	–

Thermal conductivity units

Watt/(meter Kelvin)	Kilocalorie/(hour meter centigrade)	British thermal units inch/ (hour feet square Fahrenheit)	British thermal units/ (hour feet Fahrenheit)
W/(m · K)	kcal/(h · m · °C)	BTU · in/(h · ft^2 · °F)	BTU/(h · ft · °F)
1	$8.5985 \cdot 10^{-1}$	$6.9335 \cdot 10^{0}$	$5.7779 \cdot 10^{-1}$
$1.1630 \cdot 10^{0}$	1	$8.0636 \cdot 10^{0}$	$6.7197 \cdot 10^{-1}$
$1.4423 \cdot 10^{-1}$	$1.2401 \cdot 10^{-1}$	1	$8.3333 \cdot 10^{-2}$
$1.7307 \cdot 10^{0}$	$1.4881 \cdot 10^{0}$	$1.2000 \cdot 10^{1}$	1

13 Calculations

The molar ratios of the isocyanate-reactive components and the isocyanate must be calculated for each formulation. First, the stoichiometric ratio of the two reactants is calculated. The formulator can decide to deviate from this ratio by using more or less isocyanate. The number average functionality of mixtures of reactants is required to predict the gelling behavior of the system. Hard block content calculations are required in elastomer development work. The basic equations are summarized, and some calculations are presented.

13.1 Basic equations

Symbols

EM	Equivalent mass
M	Molecular mass
f_i	Functionality of molecule i
f_n	Number average functionality
n_i	Number of molecules i
OHv	Hydroxyl value
NCO value	Isocyanate value (in wt.%)
w	Weight amount
mol	Number of moles
CE	Chain extender

Equations of hydroxyl and isocyanate values

$$(1) \quad OHv = \frac{f_n\text{-}polyol \cdot 56.1 \cdot 1{,}000}{M_{OH}} = \frac{56.1 \cdot 1{,}000}{EM_{OH}}$$

$$(2) \quad NCO\ value = \frac{f_n\text{-}iso \cdot 42 \cdot 100}{M_{NCO}} = \frac{42 \cdot 100}{EM_{NCO}}$$

Calculation of the index

$$(3) \quad Index = \frac{mol\ NCO}{mol\ OH} \cdot 100$$

https://doi.org/10.1515/9783110744583-013

The required amount of isocyanate

The required amount of isocyanate for a given recipe can be derived by transformation of equations (1) to (3):

$$(4) \quad w_{NCO} = \frac{OHv \cdot w_{OH}}{1{,}336 \cdot NCOvalue} \cdot Index$$

Functionality

$$(5) \quad f_n = \frac{\sum n_i \cdot f_i}{\sum n_i}$$

Hard block content of elastomers

$$(6) \quad HBC = \frac{100 \cdot (w_{NCO} + w_{CE})}{(w_{polyol} + w_{NCO} + w_{CE})}$$

13.2 Worked examples

Amount of isocyanate required for a given recipe

100 g of a trifunctional polyether polyol with an OH value of 420 mg KOH/g is reacted with polymeric MDI (NCO value: 31 wt.%) at an index of 110. Calculate the amount of required isocyanate.

Solution via equivalent masses

Step 1: Calculation of the polyol equivalent mass: 133.57 g/mol
Step 2: 100 g of polyol corresponds to 0.7487 equivalents.
Step 3: Calculation of the isocyanate equivalent mass: 135.48 g/mol.
Step 4: Thus, for an index of 100, the required amount of isocyanate is
135.48 · 0.7487 g = 101.43 g.
Step 5: For an index of 110, the required amount is 1.1 · 101.43 g = 111.6 g

238 —— 13 Calculations

Solution via equation (4)

Inserting the numbers into equation (4) gives

$$w_{NCO} = \frac{420 \cdot 100}{1{,}336 \cdot 31} \cdot 110 = 111.6\,g$$

13.3 Average functionality of a polyol mixture

Calculate the number average functionality of a polyol formulation that contains 90 g of a trifunctional polyether polyol with an OH value of 250 mg KOH/g and 10 g ethylene glycol.

Solution

Step 1: Calculation of the triol polyol molecular mass: 673 g/mol
       The molecular mass of ethylene glycol equals 62 g/mol
Step 2: Calculation of the molar amounts of both polyols (n_i)
Step 3: Calculation: molar amount of polyol multiplied by its functionality $(n_i f_i)$
Step 4: Calculation of the sum of $(n_i f_i)$ and sum (n_i)
Step 5: Dividing the sum of $(n_i f_i)$ by the sum of (n_i) gives the number average functionality $f_n = 2.46$

Component	M (g/mol)	f_i(OH)	Amount (g)	n_i (mol)	$f_i \cdot n_i$ (mol)
Polyether polyol	673	3	90	0.134	0.402
Ethylene glycol	62	2	10	0.161	0.323
Sum			100	0.295	0.725

13.4 Hard block content of an elastomer

100 g of a difunctional polyester with a molecular mass of 2,000 g/mol and 1,4-butanediol ($M = 90$ g/mol) are reacted with 4,4'-MDI (NCO: 33.5 wt.%) to form an elastomer at an index of 100. Calculate the amounts of 1,4-butanediol and 4,4'-MDI required to achieve a hard block content of 40 wt.%.

Solution

The required amount of 4,4′-MDI depends on the amount of chain extender. Thus, equation (6) contains two unknowns: w_{NCO} and w_{CE}. Hence, we must transform equation (6) into a form with only one unknown.

Step 1: Calculation of the OH values of the polyester and the chain extender:
$$OHv_{polyester} = 56 \text{ mg KOH/g}$$
$$OHv_{CE} = 1{,}247 \text{ mg KOH/g}$$

Step 2: Step 2: Required amounts of 4,4′-MDI to react with the polyester and chain extender:
$$w_{NCO(polyester)} = 12.53 \text{ g}$$
$$w_{NCO(CE)} = 2.79 \text{ g per gram of chain extender}$$

Step 3: Required total amount of 4,4′-MDI:
$$w_{NCO} = 12.53 + 2.79 \cdot w_{CE}$$

Step 4: Substitution gives:
$$HBC = \frac{100 \cdot (12.53 + 2.79 \cdot w_{CE} + w_{CE})}{(100 + 12.53 + 2.79 \cdot w_{CE} + w_{CE})}$$

Step 5: The amount of chain extender for an HBC = 40 wt.%:
$$w_{CE} = \frac{112.53 \cdot HBC - 1{,}253}{379 - 3.79 \cdot HBC} = \frac{112.53 \cdot 40 - 1{,}253}{379 - 3.79 \cdot 40} = \frac{3{,}248.2}{227.4} = 14.3 \text{ g}$$

Step 6: The total amount of isocyanate equals 52.4 g

Cross check

$$HBC = \frac{100 \cdot (w_{NCO} + w_{CE})}{(w_{polyol} + w_{NCO} + w_{CE})} = \frac{100 \cdot (52.4 + 14.3)}{(100 + 52.4 + 14.3)} = \frac{6.670}{166.7} = 40.0 \text{ wt.\%}$$

Index

Acidolysis 227
A-component 63
Aeration 113
AFM 70
Aging 141
Airflow 161, 179
Allophanate 53, 64, 66
Amide 51
Aniline 12
Anisotropy 136
Annealing 211
Appliance 153
Artificial leather 222
Atomic force microscopy 70
Axial piston pump 107

Ball rebound 86, 161, 179
Band casting 219
Base-phosgenation 11
B-component 63
BDMAEE 56
Bio-based polyol 35
Bis(2-dimethylaminoethyl)
    ether 56
Biuret 53, 66
Blowing agent 40
– Chlorofluorocarbon (CFC) 41
– Hydrocarbon (HC) 42
– Hydrochlorofluorocarbon (HCFC) 41
– Hydrofluorocarbon (HFC) 41
– Hydrofluoroolefin (HFO) 42
Bubble formation 116
– Aeration 113
– Nucleation 113
Bubble growth 115
Bunstock 148
Burning behavior 93
– Fire growth rate 147
– Fire triangle 93
– Smoke growth rate 147
– Total heat release 147

Capping (with EO) 29
Carbamic acid 51
Carbodiimide 17, 58
Carbon footprint 224
Cast elastomer 219

Castor oil 225
Catalyst 54
– Amine 55
– Tin 56
Cell 123
– Anisotropy 136
– Closed-cell 135
– Open-cell 159
– Size 136
– Structure 123, 135
– Strut 135
– Window 135
Cell opening 120
CFC 41
Chain extension 21
Chain topology 194
Chemical blowing 40
Chemical recycling 227
Chlorofluorocarbon 41
CLD 176
Closed-cell 135
CME-foam 184
Cold-cast elastomer 219
Cold-cure 181
Cold-hot-phosgenation 11
Collapse 121, 171
Combustion-modified ether 184
Compression hardness 161, 176
Compression load deflection 176
Compression set 161, 213
Compressive strength 144
Cone calorimeter 96
Continuous processing 98
Conventional flexible foam 183
Cream time 131
Crib 5 162
Crosslinker 37
Crude MDI 14
Crystallization (strain-induced) 215
Cyclopentane 42, 153
Cyclotrimerization 61

Day tank 104
Density 123
– Distribution 149
– Foam density 123
DETDA 38

https://doi.org/10.1515/9783110744583-014

Diethyl toluene diamine 38
Differential scanning calorimetry 89
Dimensional stability 142
Dimerization 60
Dinitrotoluene 18
Discontinuous processing 98
DMA 84, 87
DMC catalysis 27
Dome formation 148, 184
Double conveyor belt laminator 151
Double metal cyanide catalyst 27
DSC 89
Dual-hardness 181
Dynamic mechanical analysis 84, 87

EA foam 186
Elasticity 214
Elastomer 190
– Chain topology 194
– Cold-cast 219
– Fiber 222
– Hot-cast 220
– Microcellular 220
– Morphology 196
– RIM 221
– Spray 222
Elongation at break 207, 213, 215
Emission 164
End of rise time 132
Energy-absorbing foam 186
EO tipping 29
Equivalent mass 22, 24
Exotherm 51
Extrusion 218

Facing 152
Fiber 222
FIGRA 147
Filled polyol 30
– Graft polyol 30
– PHD polyol 31
– PIPA polyol 31
– PMPO polyol 30
Fire growth rate 147
Fire triangle 93
Flame retardant 44
– TCPP 44
– TEP 44

Flammability test 95
– Cone calorimeter 96
– Crib 5 162
– Limiting oxygen index (LOI) 96
– Small burner test 146
– UL 94 96
Flat-top process 184
Flexible foam 159
– Cold-cure 181
– Combustion-modified ether 184
– Conventional foam 183
– High resilience 165
– High resilience foam 183
– High-load bearing 184
– Hot-cure 182
– Hypersoft 184
– Low-density 184
– Molded foam 181
– Morphology 174
– Packaging foam 188
– Slabstock foam 183
– Slow recovery foam 187
– Viscoelastic foam 184, 187
Floating lid 148
FMVSS 302 162
Foam density 123
Foam formation 111
– Bubble growth 115
– Cell opening 120
– Foam stabilization 120
– Gibbs-Marangoni effect 121
Foam hardness 178
Foam porosity 123
Foam-in-place packaging 188
Functionality 22–23, 26, 37

Gas phase phosgenation 11
Gas thermal conductivity 140
Gear pump 106
Gel profile 172
Gel time 131
Gelation 77, 132
Gibbs-Marangoni effect 121
Glass temperature 129
Glass transition 82
Glycolysis 227
Greenhouse warming potential 41
GWP 41

H$_{12}$MDI 9, 23
HACS 162
Hard block content 166
Hard block structure 202
Hard segment 202, 205
Hardness 3, 161, 174, 177, 212
HCFC 41
HDI 9, 23
Headliner 185
Hexamethylene diisocyanate 9
HFC 41
HFO 42
High resilience foam 165, 183
High-load bearing foam 184
High-pressure mixing head 108
High-pressure processing 101
– Impingement mixing 101
– Reaction Injection Molding 102
HLB-foam 184
Hooke's law 84
Hot-cast elastomer 220
Hot-cure 182
HR foam 165
Humid aged compression set 162
Hydrocarbon 42
Hydrochlorofluorocarbon 41
Hydrofluorocarbon 41
Hydrofluoroolefin 42
Hydrogen bond 52
Hydrogenated MDI 9
Hydroxyl value 37
Hypersoft foam 184
Hysteresis 179

IMC 105
Impingement mixing 101
Index 64
Infrared spectroscopy 52
Inline system 102
In-mold coating 105
Instrument panel 185
Insulated pipe 154
Insulation board 152
Integral skin foam 220
Internal mold release 99
Interphase 196
IPDI 9, 23
IR spectroscopy 52

Isocyanate 9
– HDI (hexamethylene diisocyanate) 9, 23
– IPDI (isophorone diisocyanate) 9, 23
– MDI (methylene diphenyl diisocyanate) 9, 12, 23
– NCO-prepolymer 20
– NDI (naphthalene diisocyanate) 9, 19, 23, 89
– PDI (pentamethylene diisocyanate) 9
– PMDI (polymeric MDI) 16, 18, 23
– PPDI (p-phenylene diisocyanate) 9
– TDI (toluene diisocyanate) 9, 18, 23
– TMXDI (tetramethylxylylene diisocyanate) 9, 23
– TODI (o-tolidine diisocyanate) 9, 23
– H$_{12}$MDI (hydrogenated MDI) 9, 23
Isocyanate index 64
Isocyanurate 58, 61, 66, 127
Isopentane 42
Isophorone diisocyanate 9

k-Factor 42, 138
Kinetics 74

Lambda value 42, 138
Laminated board 150
Laplace pressure 116
LD-foam 184
Light stability 57
Limiting Oxygen Index 96
Loss factor 86, 179
Loss modulus 86
Loss tangent 86
Low-density flexible foam 184
Low-pressure mixing head 108
Low-pressure processing 100

Mannich polyol 39
Maxwell model 84
MBOCA 39
MDI 9, 12, 23
– Crude 14
– Isomer reactivity 48
– Liquified monomeric 61
– Monomeric 16
– Polymeric 16, 18
– Pure MDI 16
Metal sandwich panel 153

Metering 106
- Axial piston pump 107
- Gear pump 106
- Piston pump 106
- Plunger pump 106
Methylene diphenyl diisocyanate 9, 12
Methylene-bis(2-chloroaniline) 39
Microcellular elastomer 220
Mixing head 107
- High-pressure 108
- Low-pressure 108
Mixing time 131
Modulus 82
- Loss modulus 86
- Storage modulus 86
- Young's modulus 84, 144, 176
Mold 104
Molded foam 181
Monomeric MDI 16
Montreal Protocol 41
Morphology 196

Naphthalene diisocyanate 9, 19, 89
Natural oil polyol 36, 225
NCO value 22
NCO-prepolymer 20
NDI 9, 19, 23, 89
NOP 36, 225
Nucleation 111, 113

OCF 156
Odd-even effect 205
ODP 41
OH value 37
One-component foam 156
One-shot method 98
o-Tolidine diisocyanate 9
Overpacking 99
Oxazolidinone 54
Ozone depletion potential 41

Packaging foam 188
PDI 9
Pentamethylene diisocyanate 9
Pentane 42
Petzetakis process 184
Phase separation 208
PHD polyol 31
Phosgenation 10

- Base-phosgenation 11
- Cold-hot-phosgenation 11
- Gas phase phosgenation 11
Photooxidation 57
Physical blowing 40
PIPA polyol 31
Pipe-in-pipe insulation 154
PIR 62, 127
Piston pump 106
Planiblock process 184
Plunger pump 106
PMDI 16, 18, 23
Polycaprolactone 33
Polycarbonate polyol 34
Polyester polyol 32
Polyether amine 39
Polyether polyol 23
Polyisocyanurate 62, 127
Polymer polyol 30
Polymeric MDI 16, 18
Polyol 23
- Bio-based polyol 35
- Filled polyol 30
- Mannich polyol 39
- Natural oil polyol 36
- Polycaprolactone 33
- Polycarbonate polyol 34
- Polyester polyol 32
- Polyether amine 39
- Polyether polyol 23
- Poly(tetramethylene ether) glycol 32
Poly(tetramethylene ether) glycol 32
Porosity 123
Pounding test 162
Pour-behind technology 105
PPDI 9
p-Phenylene diisocyanate 9
Prepolymer 20
- Chain extension 21
Prepolymer process 98
Processing 98
- Continuous 98
- Discontinuous 98
- High-pressure 101
- Inline system 102
- Low-pressure 100
- One-shot method 98
- Prepolymer process 98
- Recirculating system 101

PTMEG 32
Pump 106
- Axial piston 107
- Gear 106
- Piston 106
- Plunger 106

Reaction injection molding 102
Reaction profile 131
- Cream time 131
- End of rise time 132
- Gel time 131
- Mixing time 131
- String time 131
Reactive extrusion 218
Reactivity 54
- Isocyanate 47
- Polyol 29
Recirculating system 101
Rectangular shape system 184
Recycling 226
Relaxation behavior 84
Release agent 99
Renewable resource 224
Resilience 161, 213
Reverse heat leakage 154
Rigid foam 125
- Appliance 153
- Insulated pipe 154
- Insulation board 152
- Laminated board 150
- Metal sandwich panel 153
- One-component foam 156
- Rigid slabstock 148
- SFC 156
- Spray foam 155
Rigid slabstock 148
RIM 38, 102
RIM elastomer 221
Rise profile 132
Rotary table 106

SAG factor 177
Sandwich panels 147
SAXS 70, 91
SBI test 147
Second rise 62, 133
Semi-rigid foam 185
SFC 156

Shore hardness 212
Shrinkage 143, 172
Single Burning Item 147
Slabstock foam 183
- Flat-top process 184
- Petzetakis process 184
- Planiblock process 184
- Rectangular shape system 184
- Vertifoam process 185
Slow recovery foam 187
Small burner test 146
SMOGRA 147
Smoke growth rate 147
Soft segment 195
Sound absorption 162
Spinodal phase separation 169
Spray elastomer 222
Spray foam 155
Spray foam can 156
Starter 24
Steric hindrance 38, 48, 55
Stoichiometry 64
Storage modulus 86
Strain-induced crystallization 215
String time 131
Strut 135
Support factor 177
Surfactant 43

TCPP 44
TDA 18
TDI 9, 18, 23
- Isomer reactivity 48
Tear strength 213
Technical foam 185
TEDA 55
TEM 70
Tensile strength 213
TEP 44
Tetramethylxylylene diisocyanate 9
TF foam 186
Thermal conductivity 137
- Aging 141
- Cell gas 138
- Radiation 138
- Solid 138
Thermo-formable foam 186
Thermoplastic polyurethane 218
THR 147

246 —— Index

Time-temperature-transformation 77, 132
Tin catalyst 56
Tipping (with EO) 29
TMXDI 9, 23
TODI 9, 23
Toluene diamine 18
Toluene diisocyanate 9, 18
Topology 194
TPU 218
Transmission electron microscopy 70
Triethyl phosphate 44
Triethylenediamine 55
Trimerization 59, 61, 127
– Reaction mechanism 61
Tris(2-chloropropyl) phosphate 44
TTT diagram 77, 132

UL 94 96
Underwriters Laboratory 96
Urea 51
Urethane 49
Uretidinedione 58

Vacuum insulation panel 140
VE-Foams 184
Vertifoam process 185
VIP 140
Viscoelastic foam 184, 187
Viscoelasticity 84
Viscosity 18, 24
Vitrification 73, 77, 132
VOC 164
Volatile organic compound 164

WAXS 91
Window 135

X-ray scattering 70, 91
– SAXS 91
– WAXS 91

Yellowing 57
Young's modulus 84, 144, 176